Short Stories of the Great Open Road and Life

By

E. Larry Smith

Copyright © E. Larry Smith 2016

All rights reserved

Note:

"The Copyright Act of 1976 provides that Copyrights begin at the moment the work is created. Registration with the Copyright Office is not required for the work to be protected under Copyright law."

Dedication

These stories are dedicated:

to all those warriors who have lived on the great open roads of this country and to

my children: Nicole and Kristin and to

my grandchildren: Jaden, Jamison, and Bentley

with hopes that also they find greatness.

The Open Road

*A call came to me, from the open road,
and oh, how it beckoned me.
It spoke indeed to my very soul,
just a young guy from Tennessee.*

*Dallas and Denver, and Frisco,
New Orleans, New York, and LA,
new places, new people, new stories,
great music with new songs to play.*

*The grandeur of valleys and mountains,
wonder of the awesome great southwest,
endless miles; the heartland and great plains,
everywhere gets better, and there is no best.*

*Miami, Key West, and Atlanta,
adventure found in many places;
Detroit, Cleveland, and Chicago,
12,000 miles and 10,000 faces.*

*I used my thumb to explore this land,
at least for a number of years,
and then I mastered the big rigs,
and I shifted a lot of gears.*

*Just a young lad, a young nomad
with diesel flowing in my blood,
crawling along in a blizzard,
slipping and sliding on ice and mud.*

*I grappled that wheel for twenty-three years,
all in all, over three million miles,
chasing clouds, sun, moon, and stars through
times of tribulations, and many more with smiles.
I drove produce rigs from California*

to markets in NYC and Boston.
An endless, open road of adventure;
Pittsburgh, Cleveland, Nashville, and Austin,

onions from Mexico to Philly,
furniture to ghettos in Chicago,
seafood from Mass to San Francisco,
and on and on with every kind of cargo,

Coke machines to cowboy towns in Texas,
metal rolls to the factories in Detroit,
pretty clothes to Portland and Seattle,
or any place this gypsy could exploit.

There's a song that lingers in my mind;
a haunting hum of horses under hood,
the rhythm of wheels as they kiss the road
as they roll on and on, and that's good.

Roll on through ice storms in Texas,
blizzards in Wyoming and Ohio,
bad traffic jams, storms and detours,
and the Badlands at 40 below.

Rock on; through coastal hurricanes, and
floods throughout the Mississippi Valley,
crossing deserts and Death Valley at 110,
east and west on through tornado alley.

A guy from Tennessee, now come home
who found out what life can be worth.
I have felt the heartbeat, and touched
the soul of the greatest nation on earth.

Written for and dedicated to E. Larry Smith
    by Jim Collier

Short Stories of the Great Open Road and Life

Table of Contents

Title	page no.
Big Boys - Big Rigs	1
Frisco Bound	3
The Haunted House	6
The Night I Spent a Year in a Blizzard	8
California Girl	10
Hot Coffin and a Cadillac	13
Warts and All	15
Buzzard and Mexico	17
No Walking into Canada	21
Fat Daddy vs. L&N and the Railroad	23
Drunks and Guns	24
Love and Guns	25
Quiet Ride Home	26
Motor Passion	28
Midnight Rider	30
On the Boulevard	31
Road Dog Stories	33
My Breakdown Broke Down	37
My First Car	38
Chicken Boy and More	40
New York City Phobia and More	42
My Old Saddle Pal Bob	43
Old Love Letters	45
Airplanes	46
Live Through This	47
Music	49
Reflections of Joe Copley	51

Doc Mongle and Sam	52
Cruisin'	54
Sergeant Peppers Merry Pranksters	55
Flying to Knoxville	57
Duct Tape and Onions	58
Truckin' with Earl	60
Forty Below	62
Race Weekend	64
Troubles	65
Fast Ride	66
Time	68
Coconuts and Mustard	69
High Water	71
Jess and Alma	73
John Ed Robbins and the War (A story from a different time)	75 – 81

The Big Rigs - The Big Boys

I have driven every kind of truck that's ever been built: LA to New York, Chicago to Mobile, and all points in between. Hammer down in the fast lane, day and night, the never ending ride captures a yearning in my soul to go where I have not been, to see what I have not seen.

My daddy was a trucker. I am a trucker's son. Diesel is in my blood. I'm at home anywhere I go. I have lived my life on the boulevard. I've driven produce rigs from Southern California to market in Boston and New York. I delivered furniture in the ghettos of Detroit, Coke machines to cowboy towns in West Texas clothing to Seattle, frozen seafood from Gloucester, Massachusetts to San Francisco, rolls of metal to car factories in Alabama and Michigan, onions from Mexico to food warehouses in Philly.

In a trucker's life, there are two seasons: winter and construction. I've driven in blizzards in Wyoming, tornadoes in Alabama and Mississippi, hurricanes in Florida and Louisiana, ice storms in Texas and floods in the Midwest. The desert heat is a test of people and machine. I've been in Death Valley when the air conditioning could only keep the temperatures around 100°. I've been in Minneapolis at 40° below.

There were times when I owned the rig and was a wildcat; totally on your own and getting your own loads. There were no cell phones in those days, and I spent a lot of time in phone booths in truck stops. If you stay in business six months, you can establish a relationship with shippers and learn who to trust. I've hauled loads and was never paid, been lied to and ripped off at every turn. They see you as a greenhorn and think you won't last long so they give you loads that no one else wants. After a while, you learn the ropes. Sometimes you might bribe the shipper, sometimes you must haul loads that are overweight so you learn where you must stop and get weighed, and find another route. You must sometimes be illegal to stay in business.

    I once managed a trucking company. I enjoyed the challenge and I was good at that. After a few years I grew tired of office politics and went back to the road. I drove over 3,000,000 miles with no accidents. I was lucky and retired from a union job with a good pension. There's a rhythm to the wheels as they roll on through the night; rocking and rolling across this magnificent land.

Frisco Bound 1965

John Robbins and I first hit the road in 1965 by riding our thumb from Bristol, Tennessee into San Francisco. High adventure for two small-town kids. We were looking for America. Every ride was about meeting new people and exchanging life's adventures. Getting into Frisco was tough hitchhiking so we rode in on a bus; wide eyed at the fantastic grape vineyards in the Napa Valley, and crossing the Golden Gate Bridge in a fog. Arriving at the bus station, we saw baton-wielding cops beating the homeless lying on the sidewalks. A local kid offered to show us the way to Haight-Ashbury, our final destination. While walking through alleys, I noticed he kept trying to get behind us. He came out with a knife, and I came out with an old-style can opener with a sharp point. He looked at the can opener and split. A Hell's Angel came blasting through the alley as a welcome to a San Francisco moment.

Standing on the corner of Haight-Ashbury, we bought acid that was fake. Thinking we sure were a long way from home! Man if they could only see us now; living the Kerouac dream - <u>On the Road,</u> every moment an adventure: Jack Kerouac's Dharma Bums - a religious wandering; Itchycoo Park, and The Small Faces; and the San Francisco Diggers, that guerilla street theater group who gave free food. Working in car washes and selling blood. Poetry and music in the cafés at night and revolution in the air. Santana playing for free in the streets of Palo Alto, Arthur Brown's "Fire" on the jukebox while we were eating fish at Fisherman's Wharf. Man this was the California dream; hills that go straight up with the streetcars. We had seen a lot of it in the movies, but here it was for real. All the music concerts at the Fillmore West; Jefferson Airplane, H.P. Lovecraft, Moby Grape, Grateful Dead, Janis Joplin/Big Brother and The Holding Company; the greatest voices of the time with raw minimalist, high-energy rock bands. Wow! Bell-bottomed, full-blown Hippies; purple afroed, acid-eating, patchoe oiled, mantra humming definitions of cool; leather clad Hell Bikes, louder than a train wreck, commanding all attention and leaving you in frightened amazement.

Soon summer was over and we began our journey home. On the map I-80 looked good, but it headed north. Knowing nothing about weather, we were soon caught in a blizzard, and the road was closed. Sitting in a bus station, we were approached by an Indian who promised he would have us in Tennessee by walking through the wilderness in three days. We declined, and I still wonder about his motives. He would have probably tried to kill us in the first mile.

We had to backtrack through California to get to the southern route, and eventually we found ourselves in Dallas, Texas. While standing on an exit by I-40 overlooking a railroad yard, we decided that if it was good enough for Woody Guthrie, it was good enough for us. Hobos directed us to an eastbound train, and we climbed aboard. Hot damn, riding the blinds just like ole Woody. We awoke the next morning in Houston - hundreds of miles in the wrong direction! I believe the hobos are still laughing.

We caught our last ride in Arkansas. It was a drunk man in a Chevy truck who asked me to drive, and gave me his credit card. He climbed in the back and went to sleep. We were Tennessee bound now. We left the gentleman at dawn sleeping in the Tennessee-Virginia Rest area. We walked home in opposite directions thinking about our summer adventure. I remember thinking we had crossed the Rubicon. Like Tolkien's Bilbo Baggins, like Jack Kerouac's Sal Paradise, and as in Homer's Odyssey, ---- we have been there and back again.

The Haunted House

Well, it started out as a dare. Who had the nerve to spend the night in the haunted house? Six of us, all around 12 years old, decided to take that dare. We were all going. The old house had history. Supposedly, during the Civil War some renegade soldiers had killed a young woman in this house. At night, as you might expect, she had been seen roaming the old house. We had heard this all our lives so we were sure. We had a lantern and a 22 rifle just in case. We made dinner and had a good time until everyone went to sleep around midnight. I was awakened by a loud noise downstairs. My friend was wide-eyed and frozen in a trance. His eyes stared straight ahead, but we heard nothing but the noise downstairs. Everyone woke up and were scared to death. It was a very loud noise. I had the rifle and the lantern. As I eased down the steps, I could see a cow that had somehow gotten in the house and was bumping into everything. I screamed like I was dying, and shot the gun to give my friends a thrill. When they found out I was kidding, no one laughed. They could not see the humor because of the horror. They were afraid standing in the dark. A haunted house is no place for cow- ards. Now, if I see one of my buddies that was there, we still get a laugh, and it's always the first thing we talk about.

The Night I Spent a Year in a Blizzard

I left Columbus, Ohio headed for Chicago about 10:00 PM. The broadcast media was blasting warnings about the approaching blizzard. I did not want to go, but my trucking company wanted to tell the shipper I was in route rather than sitting at the terminal. I had no choice. Within one hour of departure, there was no traffic on Interstate 70. I always had a policy that if the truck scared me, I would stop, but this rig was heavily loaded, and the weight helped to keep it moving. If you stopped, you probably wouldn't go again. Two hours into the journey, I stopped at a truck stop. Some of my friends were there. They were trying to decide whether to continue. They were going to Columbus which meant I was on my own. Interstate 70 was deserted as I made my way to Indianapolis. As I came off the ramp to get on I - 75, a salt truck was sitting in the middle of the ramp with its lights off. There was no chance of stopping or even slowing. Easing the steering wheel to the left and up on the hump of the ramp, I almost avoided a catastrophe, but the sides of my rear tires caught on the plow blade. As I glanced in my mirror, I could see the salt truck spinning around behind me. From what I could see, it spun around about four times. Some dummies sitting in a salt truck on a ramp, smoking cigarettes and laughing. I'll bet they weren't laughing after that ride. I had no place to pull over and check my rig for about 30 min. Only the tires had scrape marks. No damage at all. I was a God-fearing trucker and fearless man.

As I moved on towards Chicago, I could not stop. All of the truck stops were full according to the CB, so there was no place to stop. There was no other traffic in that blizzard, and I was just moving about 25 mi./h in the worsening storm. Worsening does not describe my nightmare. Look up blizzard in the dictionary and multiply by ten. I was as focused as one walking a high wire. as tense as a coiled spring, living with a level of anxiety like I had never experienced. Anxiety often yielded to fear. Hour after hour with no relief. Concentration to the point my eyeballs bulged. I would call it a living Hell except it was too cold. Have you ever driven 25 miles an hour without hope of going faster for hours?

When I finally reached Chicago, there was also an ice storm. The roads were slanted, and all of the trucks were in a mostly jackknife, but still creeping along. I had no plans to stop. I had to go to the terminal. The dispatcher was surprised to see me. I was the only truck that made it in. Had I stopped at any point, it would have been the end of that ride. When I left the truck, I could hardly stand. I didn't realize how tired I could be. One trip, one blizzard, one night that lasted a year. I was five hours over running time and as illegal as Al Capone.

California Girl

The big blue Ford was slowing down and headed for the shoulder of the road. I yelled at Bob to come on. I opened the passenger door, and she was crying. In the backseat, sat a German Shepherd. Bob was climbing into the back seat to face this huge dog. He gently slid in with almost no sound and sat quietly. In a minute or two, she had regained her composure and said, "My name is Sheila ". We introduced ourselves and she eased the big Ford back into traffic.

After a few miles, she began to talk. Her grandfather had recently died, and she was grieving. She lived in a suburb of LA. After an hour of sweet conversation, I was totally infatuated with her. She was a surfer girl; golden long hair, and I couldn't keep my eyes off her. After a while, I could see she was liking me more and more. This began a two-year love affair, the greatest love of my young life. With love you just throw yourself headlong. It is all consuming.

Here we were in LA in 1969, living in an adobe house on a tree-lined street. How could it get any better? Sheila had show-business friends as she was a divorcee of a CBS producer. She thought I should be a standup comedian. Although I was green, I was ready to go. I had an audition on a comedy show. I found out quickly this wasn't for me. I choked like a schoolboy. These guys were pros and they made a fool of me.

Bob stayed around for a few weeks and decided to go home. He was my hometown friend who talked me into hitchhiking to California again. We were both students and needed to get back. Bob split and I got a job at Cal Corporation spray painting helicopters bound for Vietnam. The pay was great, and we were living the American dream. Sheila was a nurse.

In LA at that time, people took having a good time seriously. On Friday mornings, the freeway would begin to fill up with motor homes, campers, and motorcycles. We were within a three-hour drive from the mountains, the oceans, the desert or the beauty of the valleys of the Pacific Coast Highway (PCH), highway #1. We spent many weekends sleeping on the beaches of Big Sur and Monterey.

LA was the center of rock 'n roll music at the time. Great artists like CSNY, the Eagles, Jackson Browne, Linda Ronstadt, and Dave Mason were defining the LA sound. We cruised Sunset Boulevard and Whittier Blvd. at night. There were two kinds of cruisers; the racers with fast cars and the low riders with hydraulics to make them dance, and many motorcycles of all description. Everyone got along, and in my two years in LA, I never saw a fight.

On the dark side, we met street people who told us Jesus had returned and they wanted us to meet "Charlie". as in Charlie Manson. Later, these people became famous as his 'family'.

I called back to Bristol to get my old rock 'n roll band out there. Unbelievably, they came. We played in and around LA and even quit our day jobs. We eventually rented a house in Burbank and were living the R&R dream. We rehearsed all day and played clubs at night. We had no illusions about being rock stars, but here we were. We all bought classic cars as the car culture in California was huge.

It all came to an end in 1971 with a big bang. The San Andreas Fault had shifted; waking us up at daybreak. The sky was blood red, and we had been shaken out of our beds. All the people were standing in the streets together in pajamas. One half block away a huge electrical station was exploding. The ground was rolling and pitching up and down. We all thought this was a big one, and we were going to die. After a day or two, things settled down and the Bristol women said, "We're going home ". Everyone left driving a classic car and pulling one.

Sheila and I lived in romantic bliss for another few months. I became homesick and decided to return. All the colors in Southern California are shades of brown, and I missed the green mountains. She was a California girl and wasn't willing to go with me. We both realized our worlds were too far apart. I told her I would return in a few weeks, but we both knew it would never happen. About a year after I returned, she came to see me. She stayed about a week, and I showed her our beautiful mountains - my Tennessee. We had a wonderful time. As I watched the big airplane lift off and head westward bound, I knew we would never meet again. I think of her warmly and of course wonder what became of her. She is part of why I'm here now. My vision of her would be as a grandmother, on the beach, smiling as she connects to her ocean as I connect to my mountains.

My life has been 99 miles of barbed wire and sweet honey. Sweet memories act as a cushion for the barbed wire. The sweet honey comes from love, family and friends

Hot Coffin and a Cadillac

There was an old derelict building which we'd seen all of our lives, passing by it every day. Our sense of curiosity at age16 proved irresistible. John Robbins and I hid my hot rod '57 Chevy behind the old building in some trees and slipped in. The door lock was gone, no doubt taken by previous vandals. Inside, the sunlight was beaming through broken windows into a gymnasium size area. Being the avant-garde artists that we were, we always had an 8mm camera in the car. After retrieving it, we made a fire on the floor and filmed the smoke as it rose through the light of the evening sun; very heavy imagery indeed for 1964.

Upon exploring the rest of the building, we discovered covered in dust an old 1930s truck and an automobile. It was in this area that we found it; a very old, wooden, empty. small child 's coffin. We were not thieves per se, but this building hadn't been used in decades, so we liberated said coffin and drove away with it sticking out of the trunk.

My mother was sweeping the porch as we drove into the driveway. When she saw the coffin, she hit me with the broom. She even gave my friend John a lick or two telling us to get that thing away from her house not to bring it back. It was getting dark and being Friday night, we cruised around town for a couple of hours with the coffin in the trunk; our attempt to make some front page teenager news. Since we were not allowed to keep it at home there was only one thing to do; sell it to our friend Mike. Mike's dad was very wealthy and his mom had died. Being the only child, he always had money. We sold it to him for $10.00.

He put a mannequin dressed as a witch in it and kept it in his bedroom. I sure hope the statute of limitations has expired.

Later that summer, I bought the 1951 Cadillac hearse. Only 20,000 miles on the odometer, and I talked him down to $300. I wanted to haul band equipment to gigs. I mean it was like new; a 1951 Cadillac. Again, my mother told me I couldn't park it within a mile of her house. I took that Cadillac back to the dealer and explained my situation. Being a good guy, and the fact that he had sold so cheap, he gave the money back while having a good laugh. Sometimes, we have to give up what seems to be right.

So it goes.

Warts and All

Lisa lived back in the hollow far away from everything. The mountains were so steep and the hollow so narrow, she rarely saw the sun. At least, it was a treat when the sun appeared for more than a couple hours. Completely isolated, she only saw her parents and her siblings. Her life was work. Her life was one of survival; planting and tending the crops, the livestock, and household chores. She had no time to consider if she was happy. She didn't understand the concept of happiness at least as we know it today. There were no thoughts of the world outside of the hollow.

Lisa 's dad had a sawmill to supplement the family income sawing timber to sell and sawing trees for others. In the spring of 1927, Bill Mullins and his son Jacob brought timber to be sawed into boards to build his new house. His wife had died the previous year and he had seven kids to raise. He wanted to build a new house on flatter land that his father had given him.

When Lisa saw Bill Mullins and Jacob driving the wagon up the hill, she had no idea her life was about to change. She stared at the wagon driving by her house and made eye contact with Jacob. He was about the same age and smiled warmly at the girl. She avoided his glance and went back to working at the spinning wheel. She thought no more about it.

Bill Mullin's boards were cut that day and he left to begin building his new house in Dorchester. Bill and Jacob's world was not so isolated as Dorchester was a hub of business activity. Coal trains were loaded there and several new businesses were opening. Bill owned a coal tipple and was doing well.

During the summer of 1927, Bill Mullins returned to the hollow near Gobbler's Knob. His pretense was to arrange to have more lumber cut, but he really had a more sinister motive. He was looking for a wife and in particular, Lisa.

Devil John Wright has stopped his saw blade and watched the Model T slowly make its way down into the hollow. He recognized Bill Mullins and his suspicions relaxed. Devil John lit a cigarette as the car rolled to a stop.

"Good morning Mr. Wright." said Bill Mullins. "I'm here to see when you can cut some more lumber."

Devil John like doing business with Mr. Mullins as he always paid in cash when the last board was cut, and he always bought some moonshine. "I could start Monday and finish by Wednesday." he replied.

"That will do, and I need to talk about a business deal." said Mullins. "You know the town is getting bigger and people are goin' to need more lumber. My sons will cut the trees and bring them here. We can make a lot of money."

Devil John was quiet for a few minutes. "Sure, I reckon that'll be alright."

"Great. Devil John, I'll be honest with you. I'm looking for a wife. I'm interested in your daughter, Lisa."

Devil John thought about his young daughter. He knew that she was a beautiful girl and as a companion, she would be worth a lot. Devil John was not above arranging some sort of deal. Times were hard and this was certainly not unheard of. He needed Lisa to work at home so he could not consider a small price.

Bill Mullins was looking for someone to slave away raising his seven kids. The fact that Lisa was beautiful was only icing on the cake.

"What kind of deal are we talking about?" asked Devil John. "Lisa is worth a lot to me."

"I know it is hard to figure out money paid for your daughter, so perhaps you would be interested in a business arrangement."

Devil John rolled up another cigarette, took a puff or two and looked at Bill Mullins. "What you got in mind?" he asked.

Quickly Bill Mullins said, "You got a small sawmill and I got lumber. We can make a lot of money. I own several properties in Needmore, and we need to be building. I tell you we cannot lose. How about 10% of it?"

What did he say? What did he do? What happened to Lisa?

This all happened in 1927. It is really hard to belief. It is a long story that is becoming a book; <u>Poverty to Power.</u>

Buzzard and Mexico

My friend Buzz had recently completed his duty in the Air Force, and he wanted some adventure. At the time, I was an independent trucker and owned my own rig. Buzz decided trucking was for him so I took him to California and back. We left Knoxville, Tennessee with a load of T-shirts bound for LA. After a few hours, I had him driving the rig. On Sunday morning, we were in El Paso, Texas. Because we had plenty of time, we decided to cross over to Juarez, Mexico. We called a cab from the truck stop to take us over. In a few minutes, a 1961 Chevy pulled up. The car had no windows not even a windshield in the front or back. At the border, we had to leave the cab and walk through customs. All they ask us was if we were American. The cab driver drove through a tunnel and met us at the end of our walk. We had repeatedly told the driver that we wanted to go to the marketplace to buy foodstuffs and spices. We soon found out that there is a rule with cab drivers in Mexico; if your passengers are men always take them to "Boy's Town" which is filled with brothels. A man opened the cab door to escort us in. I asked the cab driver what was going on, and he told me to go in and drink a beer, and then they would leave us alone.

The building was huge and looked like it was built in the 1700s. It was obviously built for this purpose as there were bars and bedrooms everywhere. About a dozen women came to us as soon as we walked in. Buzz and I drank a beer and the women soon figured we weren't interested and left us alone. Well, almost, but that is another story.

We left that establishment and headed for the old town. We drank beer in a bar that was a hollowed out tree. There were very old pictures on the walls of old banditos; mean looking men with guns and ammunition belts across their shoulders. The bartender told us not to drink too much as the Mexican police loved Americans as they could be arrested and held for ransom.

In the marketplace, we had a great time eating local food and buying leather goods and spices. We met our cab driver, and he told us it was his job to take us to Boys Town. Buzz ask him to get him some cocaine. I was surprised Buzz would want it as he had never done any before. The cab driver asked for a $20 bill. Buzz would not give it to him so he asked to tear the $20 bill in half some howl assuring he would come back. We waited three hours and he didn't return. We met a man trying to sell phony watches. He wanted the one half of the $20 bill and said it was worth $10.00 in Mexico.

    Later that day, we saw the cab driver. He came running up to us. The best defense is offense so he acted irate, and said he had been looking for us all day. Buzz gave him the other half of the $20.00 bill back and took us to customs. What he gave Buzz for cocaine, we learned was just aspirin. He drove us to the truck stop, and we continued to LA.

Chapter 2 Mexico and Beyond

    It is a long way from El Paso to LA. Although only two days, it seemed like weeks. The Interstate 10 route is a bit longer, but it is flatter. Route 40 takes long pulls going up mountains which is hard on the truck. We arrived in East LA during a race riot. Some of the buildings were still smoking. Fortunately, my delivery was to a Korean business. None of the Korean buildings were burnt as they are known to kill you. During the riot, they stood guard with AK-47s. These were the riots resulting from the police treatment of Rodney King.

    After the Korean customers looked over the T-shirts, they decided they didn't want them. Poor quality was the issue. Sometimes this happens as the buyer is dealing for a cheaper price. They argued all day. A deal was finally reached, and we were on our way to the San Joaquin Valley to load produce.

    The San Joaquin Valley is one of the prettiest places on earth. Rich farmland, vegetable fields, orchards, vineyards-everything is grown here. On this particular trip, we were hauling for Dole foods, and we had to go to five locations to fill the rig. I didn't mind as I was being paid well and my truck likes lighter loads. The fields were filled with Mexican people. There were green immigration police cars everywhere. The places we visited was a processing plant. They were huge facilities. When you arrive, you are given a number on your windshield, and a Jeep escorted you to a parking place. When they were ready to load you, they came and woke you up. They had state-of-the-art food trucks serving great food cheaply. Food trucks are big deal in California. My rig had two bunks, and so both of us could sleep. We were awakened at 3:00 AM to follow a Jeep to the loading ramp. All of these people were Hispanic or Vietnamese, and I never knew one that I did not like. They would always put an extra crate of produce for me.

Our last stop was Chinchilla, Mexico. We had to wait but had no problems with customs. Soon we were in the good old USA and headed home.

Our delivery was Boston, Massachusetts, but we traveled across I – 40 anyway. In Knoxville, Buzz had enough of trucking. I asked the girl I had been seeing if she wanted to go to Boston and she said no. In 30 min., her sister called and said she wanted to go, and I sure did take her.

Her name was Judy and she was an illustrator for an ad agency; a very smart girl I had known for years. To her, this was going to be an adventure and she wanted some. We arrived in Boston on Saturday morning and unloaded. Our next load was fish from Gloucester, Massachusetts to San Francisco, California. We weren't expected for a couple days so we stayed in Boston. What a great town. We saw everything there. We had meals at the "No-name Café" where Billy Joel was playing. We stayed about four hours at old Ironsides and USS Constitution. there where our country was born.

On Tuesday, we loaded off a ship in Gloucester, Massachusetts; 50,000 pounds of frozen fish. The weather was very cold and snowing, and I was anxious to go. Driving all night, we missed the big nor'easter and were on our way. It's easy money to San Francisco. Great trip. I kept wondering how money was made bringing the fish across country to another ocean. I never ask themand the checks always cashed.

After unloading in Frisco and spending a few days there, we decided to go. This time the produce was all loaded in one place so we left the City-By-the-Bay. Being totally worn out and road weary, we eventually arrived in Knoxville, Tennessee. Judy had enough of trucking. Buzzard showed up with his uncle and wanted advice on buying a truck and driving it themselves. As they were greenhorns, and I knew what they were going to have to do, I totally advised them against it.

The next day buzzard showed up at my house with a brand-new Triumph Bonneville Motorcycle. He told me he would be back in the morning, and I was to ride all day, which I did. A great ride through the mountain's. He came to get his bike the next morning. We hugged, and I never saw him again.

No Walking into Canada

I once had a friend who refused to go to war. He was no coward. He truly believed he could not participate. He and his family moved to Ontario, Canada. I decided to go see them. A friend took me to Abingdon and I began thumbing a ride. The first ride took me to Roanoke. Everything went smoothly and I reached the end of I-81 in New York in a town named Wellesley Island near Fisher's Landing, near the Canadian border. There is a bridge several miles long, and I had to walk to the border. The border guards thought I was a criminal or dodging the draft. They took me into a room and questioned me for an hour and checked everything about me. They told me I was a vagrant and was not allowed into Canada. I had over $200.00 and I was no vagrant. They told me I could not walk into Canada and they turned me away. Then I had to go through American Customs. Same deal, they thought that I was a criminal, so they questioned me for an hour and finally let me go.

    I had to walk back across the bridge several miles and caught ride into Fisher's Landing. I went to the bus station and bought a ticket to get into Canada. When the bus pulled into the Canadian Customs, the same two guards boarded the bus. They recognized me and took me from the bus. Once more, I was harassed, and I thought I was going to be arrested. With much ado, they sent me back through American Customs. They treated me badly once again.

    I walked back across the bridge. This process had taken the entire day. I was exhausted, so I slept in my sleeping bag hiding in a huge water pipe in the side of the mountain. The view was so beautiful. After breakfast I headed back to the highway. Everything went well and soon I was back in Bristol, Virginia. I had a new VW bus that I left with two VI University girls because I wanted to hitchhike into Canada. You meet so many nice people this way and hear about their lives and adventures. I learned a lot about people. Some people called me a Road Dog Warrior. Altogether, I probably hitchhiked about 12,000 miles.

During the time I was in the Navy, I thought about hitting the road again so when I got home, I bought a sleeping bag and camping gear and roamed America and Mexico.

I learned you cannot walk into Canada. The only names which have been changed have been changed to protect the guilty.

Fat Daddy vs. the L&N and the Railroad

For several years, I drove tractor trailer rigs to Meridian, Mississippi and back. While I was there, I stayed at a Holiday Inn. A van would take us to the motel and back. And old gentleman named Maurice owned the service and usually drove the van. One afternoon, his uncle, "Fat Daddy" drove the van. He was 100 years old and didn't drive very well. Some railroad workers were in the van. He decided to drop them off at the railroad yard. After the guys got out, Fat Daddy turned the van onto the railroad tracks as if it were the road. I was the only passenger, and I was in the back seat -three rows away. The van was literally jumping up and down throwing me from the ceiling to the seat. Several times, I crashed my head into the ceiling of the truck. I could see ahead where two tracks crossed, and I braced myself for the crash. The van went airborne and became hung up on the tracks and stopped with the wheels suspended in midair. I was still hyperventilating as I crawled from the van and sat on the side of the hill.

One of the first responders was a railroad guy in a pickup truck. Altogether about 8 to 10 vehicles showed up including the Mississippi Highway Patrol. Everyone got out of their cars and looked at me and said the very same exact words," How the hell did this happen?".

Trains had to be stopped. It was a big deal. There was a lot of confusion and cussing and laughing. Fat Daddy didn't know what was going on as he kept shifting the gears from forward to reverse with the accelerator to the floor. Of course, this completely destroyed the gears and the motor. He mumbled a lot of things I can't repeat, mostly because I could not understand him. I'm sure some of them were not politically correct. They had to bring a crane to remove the van. The Highway Patrol took me to the Holiday Inn. Maurice never let Fat Daddy drive again.

Drunks and Guns

It wasn't easy playing in a rock 'n roll band in the 60's. Of course, we had long hair and almost had to fight everywhere we went. There were in East Tennessee many rednecks who wanted to kill hippies. We played in Southwest Virginia coal field town's; places like Grant, Crumbs Cove, Haysi and every club in between. Sometimes, redneck boys would want to fight. My piano had metal legs, four of them, just enough to save us. They knew they could rush up and take us down, but the cost would be high with broken bones We were that wild band that smoked, drank and would fight if we had to. The Coal Field people called us them "Bristol Beatles ". I think most of the redneck troubles were about girls, as we could take their girls. We played in a club, the Pink

Room where Bob Seger played. We played in buildings that were on stilts on the side of a mountain. One club owner told us not to play the <u>Wildwood Flower</u> because the people would stamp their feet so hard that the building would fall off the mountain.
    A lot of guys got drunk. Some of them had guns. Sometimes we heard gunfire outside. In one club a man shot his girlfriend in the head and killed her. He looked down at her lying there and then shot himself in the head. The bodies were on the floor lying across each other. Everyone in the club ran to their cars and were gone. The place cleared out like a tornado came through. We couldn't leave because of the equipment so we had to watch it all. It looked like gallons of blood on the dance floor. Cops were all over the place. We had to answer a lot of questions. We weren't allowed to get our equipment for a few days. Of course, this was the most violent thing I have ever witnessed, but I had seen other people shot. Drunks and Guns – Bad – Bad – Bad

Love and Guns

    Rookie and I arrived in Mississippi shortly after sunrise. We were taken to the Holiday Inn. Before we went to our room, we were standing outside smoking cigarettes. A young black woman was knocking on the door a few feet away. The door opened and a man began arguing with her. Apparently there was a woman in the room who was his lover, and his wife had caught him. After much cursing, the wife brought out a small pistol and began firing. The man's lover turned to go back into the room and the wife shot her in the buttocks. I turned to Rookie and said, "That's got to hurt" The man ran into the hotel room, and we knew he was going for his gun. The wife had run to her car, and was screaming through the parking on two wheels. I could see her mother in the front seat bent over with her hands holding her head. The man began firing at the car as it raced by. He hit the car several times, but as he fired, he was also shooting a row of expensive cars across the parking lot. I saw windows shattered in a new Lexus and several other vehicles. The man ran to a group of trees beyond the motel and hid his gun. When the police arrived about 30 people told them this and they found it. My friend Rookie went to the police and told them everything he saw. I went to my room. Rookie had to go back to Mississippi three times for the trial.

Love and guns: a dangerous combination.

Quiet Ride Home

I once played music in Richmond Virginia at the Thomas Jefferson Hotel where the inside scenes of "Gone with the Wind" were filmed. It was for the Old South University alumni. The limousines pulled up one after another. The doorman would open the door to a beautiful woman dressed in a Scarlet O'Hara gown. The men were dressed as Confederate Civil War officers. I think everyone was a general from the uniforms, sabers and feathers. It was a wonderful evening, everyone was smiling and happy.

My band was to open the ball with a big-time heavy metal band to follow. The buffet tables were packed with exotic foods. Not a penny had been spared. During the sound checks, we played a Credence Clearwater song. Afterwards, we heard one of the generals say we were a country band which didn't seem to be a good thing so when we hit the stage, we played a rock song from R.E.M. The place came apart. The generals were all on their backs doing some kind of wiggle dance and all the women were standing in a circle around them. These people were ready to rock 'n roll, and we rocked them. What a great feeling when things go right. We played for an hour and a half and left the stage. As we were leaving, the bass player for the other band said to me, "You guys rocked and now we will run them off."

    They were incredibly loud. The song was to the tune of "My Girl" by "The Temptations", but they played it Heavy Metal style which is mostly shouting and jumping around. Also, instead of the My Girl, they shouted, My Bitch which offended everyone. By the end of the first song, everyone was gone. I was standing near the door and people were asking me if we were playing again. When I said no, they said, "To hell with this.", and they all left. The ball was over and people were sweeping up. The generals and their wives were having their pictures taken coming down the same stairs that Vivian Leigh had walked down.

    Afterwards, the problems began. There is always a glitch. The promoter had promised us rooms in the big hotel. There were no rooms. He told me he had a huge house in the country, and we were sleeping there after the party. I was angry because this was not part of our agreement. I didn't have time to get a contract signed. The rest of the band wanted to go so we drove around in the country for 45 minutes. It was no mansion. just a three-bedroom suburban house. The members of his band were cooking barbecue for the party that evening where we were supposed to play again. It turned out the promoter owned the band. He flew them on private aircraft, bought their clothes, and fed them, but he didn't pay them a penny. These guys were doing it for nothing.

I was amazed. There was no party and all the musicians were lying on the floor to sleep. I told the promoter that I didn't sleep on floors and I was headed to Bristol. The agreement was to play two shows for $1000. He asked how much he owed, and I said $1000. To my surprise, he paid me, and asked if I would play more shows for him. I said no and yelled at the boys that my car was leaving for Bristol. Everyone was mad at me, and it was a quiet ride home

Motor Passion

I have always loved motorcycles. I sold my last motorcycle less than a year ago. My first was a moped in 1958. I had a newspaper route and my parents allowed me this bike. My dad always loved motorcycles so that really helped. In those days, there wasn't much chance of being stopped by police. I had no license or tags. As I got older, I had bigger and faster motorcycles. I was into drag racing, and raced a car and bike. I had a burning need for speed and bought as much speed as I could afford.

In the 60s it was an English-Triumph, Norton, BSA, Greeves and American Harley-Davidsons. In the 70s, it was Honda. The old ways of making them had advanced. English bikes were out dated. They leaked oil and were not reliable. Old world electrics(Lucas) were trouble. You couldn't start them in rain and had to keep the motor revved up to have a headlight. But they were wonderful. They felt like a real motorcycle, heavy and powerful. They vibrated like a helicopter keeping your hands numb, and I loved the look of these motorcycles. In the 70s, the Honda 750 took over. It was faster, didn't leak oil or breakdown. No maintenance and at a lower price. This ended an era of European bikes. Honda had modified his bicycle with a small motor after World War II and even made a fuel from rice. From this, Honda motors began.

My burning need for speed became exotic Japanese and Italian bikes through the seventies and eighties. Ducati, made in Italy is a beautiful, light weight bike with a nuclear motor and are the most fun to ride. They required the 'dedicated position' which is very uncomfortable with your rear end up in the air, head first, and the weight of everything on your wrists. After about 30 minutes, I had to stop and rest, but the handling was incredible. I bought DVD's and magazines and learned to ride very fast.

There were five old guys my age that bought super bikes. We were in our fifties. We raced on Highway 421 to Mountain City. Riding fast is very technical. You must understand the physics of what is going on. For instance, if the turn is left, you must steer to the right. This makes the motor bike lean over in the turn. I bought a drag bike, Yamaha V- Max, 170 horsepower on 580 pounds of motorbike. I would never be able to put into words what it feels like to go 140 miles an hour in ten seconds or explain why I would still do it today. I have ridden 170 miles per hour on two wheels. This somehow makes me feel more alive when you are on the edge. When you are on the edge - stretching the envelope. I was only hurt once, broke five bones, stayed off the motor bikes for one year and then bought the fastest one made; a Ducati that would go 200 miles per hour.

I no longer want motor bikes. I am getting older and broken bones would be a long recovery. I have driven cars 150 miles per hour, and I felt safer at a 170 on a bike. It's all about the weight of the car. The bike is like riding the horse and the car is like driving a wagon.

Midnight Rider

My friend Larry Cox and I were sitting around the house bored to death. This was in the early 70s. About midnight, we decided to go for a ride in the country. We were riding along and I noticed headlights quickly approaching in my rear view mirror. When the car reached us, it slammed into the rear of our car almost wrecking us. We couldn't believe it. Again, the car slammed into the rear of our car. I slammed the accelerator to the floor, and it was on. We were at a big disadvantage as my car was a new Pinto wagon, one of the most unraceable cars ever built. By sheer terror, I managed to get a few hundred yards ahead. We came to a sharp turn, and I slid the car around like a moonshine runner. We met the car, and it was some red neck boys; very drunk and having fun.

We crossed Norris Dam at 80 mi./h and there sat a cop. Of course he pulled us over and approached with his gun in hand. We had to lay on the ground. While lying there with a gun pointed at us, I saw the redneck boys go by. I explained to the cop those boys were slamming into our car, and we were trying to get away. He looked at the bent bumper on the brand-new car, and I guess he believed us. He was still very angry, but he let us go.

We decided to go home ASAP. There was a small strip mall by my house that had soft drink machines. We got out and were drinking a cola to calm down. I looked around and we were standing in front of a jewelry store that had been broken into, and the cops hadn't arrived. Glass was everywhere and there was jewelry on the sidewalk. Neither of us drew a breath until we were a block or two away. Arriving at my house we were too stunned to even talk about what happened for an hour or so. Sometimes, being bored to death isn't such a bad thing. I got under my bed and did not come out for three days.

On the Boulevard

I was driving a double trailer rig across Christiansburg Mountain Virginia about 3 A.M. I looked in the rear view mirror, and all I could see was fire. I stopped and got my small fire extinguisher and ran back to the rear trailer. Somehow, the grease had leaked out of the axles and turned them red hot which caught the tires on fire. They were aflame. Flames were everywhere and I knew I must get the lever pulled to release the rear trailer. I sprayed my extinguisher and while the flames were gone, I had only a few seconds to pull the pin. On my first try I did it. The flames returned quickly, and I unhooked the trailer wires and hoses. I jacked the wheels down on the trailer and ran to the truck. I pulled the trailer and rear axle away from the rear trailer. The axle was still hooked to the front trailer. Using all the extinguisher left allowed me to unhook and pull the first trailer away.

Now the axle was unhooked but was still burning. Two other drivers had stopped and emptied their fire extinguishers on the fire, but it did no good. The tires were going to explode, so we moved back. The axle weighed 3000 pounds, and it jumped 4 or 5 feet in the air as each tire exploded. It wasn't in the roadway so we let it burn itself out.

This was before cell phones and the fire was out in 15 minutes. No fire department, but soon another trucker from my company called in the report, and in a few hours a wrecker truck arrived with a new axle, and I hit the road again once I was on the Beltway around Washington DC, and the sun was rising.

It was like I was bewitched. The rear axle on the back trailer came off. I mean the rear trailer slammed into the pavement, and the truck was out of control. I went from one guard rail to the other, but hit no vehicles. This happened twice and I managed to get it stopped on the side of the road. The axle hit a pickup truck, but the driver said he didn't care and took off. He must have stolen the truck because there was a lot of damage. The axle was in the median strip about 40 feet down. They wrenched it up. I unhooked from the rear trailer and finished my run.

Another time, I was following a truck near Roanoke, Virginia near midnight. the driver of the truck went to sleep and drove off the highway. There was a highway below and he went airborne and smashed into a concrete wall. His cargo was brick's, and they were in huge piles around the truck. I arrived at the same time as a UPS driver. We had fire extinguishers in hand, but there was no chance for the driver. We could not get to him, but it didn't matter. He was killed instantly, no doubt by the crash. The fire started out as a small blue flame that we tried to put out, but it turned into an inferno. The poor guy was killed and his body was baked under the bricks.

Once a friend told me about a car driving into his rear trailer and getting stuck there. It was at night, and he didn't realize it was there. Finally, some cars started flashing their lights, so he decided to have a look. He cut the steering wheel sharply and swung the rear trailer around and he could see the lights. He stopped the rig and ran back to the car. He said the driver had a death grip on the steering wheel and could not turn it loose. He asked the driver if he was doing alright, and he said, "I was okay until you cut that Dido ". My friend told me that he had no idea what that meant. He had pulled the car eleven miles with the car brakes locked up.

The most moving moment of sadness I have ever experienced was in Birmingham, Alabama. A rusted old camper had crashed and a family of four had been killed. All four bodies were lined up at the side of the road; momma and daddy and two kids – all dead with all of their sparse earthly belongings scattered around them. Any hope that existed within that overwhelming poverty was destroyed and taken as was their lives. I was thinking about old Tom Joad- Grapes of Wrath.

That's life on the boulevard --- and death on the boulevard.

Road Dog Stories

I knew that Riverside, California was the jumping off place. After that - nothing. If I got a ride, it would be a long one. I was leaving for Boulder, Colorado through Las Vegas and Salt Lake City. Several people were waiting on the ramp where I was beginning to hitch a ride. In California then, you could hitchhike on the entrance ramp to the freeway. I politely waited until the ones that were there first had been able to get a ride. It's the code of the road.

My turn came about an hour later. In the meantime, I was talking to a hippie girl who had a huge leather bag on her shoulder. She wanted to hitchhike with me as she felt safer. I said okay, why not? It was only a few hours to Las Vegas, and I would be glad to see it again.

I remember a tractor trailer stopped and two truck drivers welcomed us aboard. I drifted off to sleep, and I was awakened by cursing and screaming. The guys had made their attempts to get sex, and the girl had a knife to one 's throat. I jumped from the rig and untied my backpack from their truck as they roared off. They left us in the middle of the desert. Nothing but Gila monsters, snakes, and the highway that stretched to both horizons. Not a car in sight, and it was very hot. I had been in worse situations, but I couldn't remember when. In about a half hour, an old Buick stopped. They were four very drunk people who were dressed in suits and evening gowns. They were party animals from the 50s. They were so drunk that we told them we didn't need a ride. There we were; stranded in the desert.

Eventually, we arrived in Las Vegas after dark. My fellow road pal called a number and a white, boss hog, El Dorado convertible with Leopard skin seats showed up. The driver was a young black man. As it turned out, the girl had 6 pounds of pot in the leather bag. After the business was over, it was our turn to party in Old Vegas. This was the old Vegas of Frank Sinatra and the Rat Pack. The casinos were downtown and none had doors. You just walked off the sidewalk. The black man was a card dealer in the casino. He was also a pimp with 15 prostitutes. He really liked me as he thought it cool that I was a road dog passing through. He took us to dinner which he paid for. He was playing

Keno and giving me the winnings. By the end of the meal, it amounted to $350, a huge amount in those days.

    We partied until dawn. I remember the sun was coming up and a fresh band was taking the stage. My friends dropped me off on the interstate ramp at sunrise. I remember riding by Salt Lake City, and it was beautiful. My map showed the interstate going many miles north to hit the eastbound interstate. There was a small road that turned east to Boulder Colorado. I chose the small road and soon there was no traffic. It was getting dark so I gave up thumbing, built a fire, and ate beans. I was playing my guitar when I heard some people calling, "Come on hippie. We like hippies.". An old Chevy full of Indians had stopped. I climbed in with two guys and three pretty girls. After a few minutes, I saw a tavern, and I ask them to stop. I carried back two cases of beer. I made friends for life. They took me to their house on the reservation. I stayed a few days with them. Everyone went into town at night to fight the white man. They were putting guns and knives in their boots. They asked me to go and I said I'm good. I'll stay here and look after the women. They laughed and in the morning some did not return. I suppose they were in jail. One night, they awakened me as they had found me a ride; an Indian who was a Marine. He was going to his base. He had a new, very fast Firebird, and he drove 120 miles an hour.

    My next ride was a new Oldsmobile with two young guys. As this turned out, they had stolen the car in New York City, and were riding around the country on stolen credit cards. We were hungry so we got off at a small town and found a restaurant. After a few minutes, we realized we were not getting service. A group of rednecks were going to beat us up. One was a miner, and on his hard hat he had written, "Hippie Killer". We quickly got back into the car and got back on the road.

Driving through Colorado, the mountains were incredible. The guys decided to stop and sleep for a period. I walked for a few minutes and a VW bus pulled over. I looked at the driver, and he was some red- head afro kid who was tripping on acid and was totally freaked out. Without a word, I went to the driver's door. He scooted over, and I drove. He went to the back and laid down. After a couple of hours, he came back to the front. He told me I had saved his life, and he took me 150 miles out of his way to the address I was looking for in Boulder.

My old friend, Steve; had written me at my address in Los Angeles. He was telling me that he had hit the jackpot in Boulder, and I needed to come. He was living in a big-time condo, had three motorcycles and two convertibles. He had two beautiful girls living with him. He was in the drug business selling kilos of pot. He told me about famous rock 'n roll stars he had sold to.

This was in the summer, and the University of Colorado was shut down. I got a room in a fraternity house for $12 a week. It was a real castle with a moat around it, and I was the only one living there.

Summer stock theater was beginning in the University Amphitheater, and I bought a season ticket. Two months of live Shakespeare plays. I decided I wanted to be an actor, and I ate it up. Each night I saw live plays; six nights a week. I remember some actors were there like Ozzie Davis. My friend would show up every day in his fancy cars showing me everything. One night, he took me into the wild country to introduce me to his friends. It was a house full of pot. Thousands of pounds I suppose. They had automatic weapons. After explaining to my idiot friend that I felt my life was in danger, I left Colorado the next day. After a few hours, I caught a ride with a farmer in a new Ford. His sons had run off, and he was going to get them when the police caught them. They were from Morristown, Tennessee. I drove most of the way. Another ride and I was home telling my friends of the adventure. I went back to school and began playing music in bars again.

    My friend in Boulder stayed in the drug business. He was busted by the police. He ratted his partners "so I heard', and was killed in prison.

My Breakdown Broke Down

    My trucking job with the union required me to pick up and deliver trailers in darkened urban terminals in the middle of the night. As with most drivers, I had a gun; a small 32 automatic. I stashed it away in the bottom of my rucksack most of the time, but there were times it was in my pocket. You hear about some driver killed by desperate homeless people, and you get a gun. The gun laws are tough. You can't take one across state lines, and I cross state lines every day. If you get caught with one, you are in deep trouble, but there were times when you needed one. I'm not going to let anyone kill me.

    Once I was coming back to New Orleans and my truck broke down in Atlanta. The company brought me another on a wrecker. It was raining, and in the confusion, I left my bag in the broken truck which was being towed to Chattanooga. I would be fired if the gun was found and maybe go to jail. I had to get that bag. I drove nonstop to Chattanooga, and could not find the broken truck.

I was told by the dispatcher that the wrecker had broken down. This was a very stressful situation, but in a couple of hours the wrecker showed up, and I got my bag. Walking into a dark truck terminal scares me. Some terminals have no fences and homeless or robbers sleep in the empty trailers. It must be the same feeling as a cop approaching a car he has stopped. There are times when you need a gun. There are a lot of desperate people in this world, and they want whatever you have, even your life. If someone has to die, better them than me. I'm for gun control, and I feel there are too many guns, but I can tell you there are times you need one as a truck driver.

I would rather have 12 trying me than six people carrying me. I never had to shoot, but I did show it to people a few times. Sometimes drivers would shoot their guns to let anyone around know they were armed. So it goes.

My First Car

Cars are everything to a boy; even one that doesn't run. A kid is pleased to sit in it for hours, pretending they are driving. My first auto was a 1953 Chevrolet Bel Air four-door with a bad motor. I was 12 years old. I got it for $75, and drove it back to my house. I'm sure I killed every mosquito in Sullivan County. It smoked so badly, cars behind me couldn't see.

I bought a motor rebuild kit from Sears and Roebuck's for $89. It had all the bearings, gaskets, rings and everything I needed, even an oil pump. I bought a Chilton Manual which gives pictures and instructions for every step. I tried to remove the motor with my sisters swing set. Of course, it broke. She has never forgiven me. When my father realized the garage was full of my car and rusty old parts, he was angry, but he eventually helped me rebuild it.

It took all winter, and I had dozens of parts left over that I had no idea of where they fit, but it started and ran well which surprised everyone including me. We lived in the country, and I could only drive it around the block, but there it was.

There was a drive-in theater a couple of miles away, and my buddies, and I spent a lot of time there. When we put the motor back in the car, we had no hoist to hold it. As kids will do, we improvised. There was a huge tree near the main highway with a limb over the road. My buddy had a 1950 Ford so we took the trunk lid off and put the motor in it. We hooked my car behind. We climbed a tree and put the chains around the limb when there was no traffic in sight. We pulled under it and hoisted the motor into the rear car. This was old highway 11 when there was no interstate, big trucks and lots of traffic.

We didn't know it couldn't be done, so we did it. I kept that old car and drove it to high school. Eventually, I bought a 1957 Chevrolet, and built a hot rod. I gave the old car to my dear friend, and he drove it for a couple years. I was told later that motor was running a pump at a friend's farm. I have had a lot of cars, but in a lot of ways, that was my favorite.

Chicken Boy and More

I went to the Tennessee Employment Office, told them I had a license to drive a tractor-trailer and boy, did they have a job for me. I became the new chicken boy. I drove big rigs at night to the chicken factory to get the chickens. Imagine 1000 chickens hanging from the conveyor belt as they were processed. I can barely eat chicken now. By day, I distributed the chickens to grocery stores in a small truck. When I arrived, the butchers would yell, "Chicken Boy is here. "Enough said about that.

Once in my life, at some point, I was a security guard. They would send me to textile factories and sweatshops where the workers were on strike. I remember I didn't last long because I was on the side of the strikers wanting a living wage.

I once repossessed furniture while in college. My job was to drive a big truck with the guy repossessing the furniture. Once the door opened a few inches and the barrel of a gun came out.

I was also a bartender listening to every hard luck story, but I was quick to light your smokes and laugh at your jokes. The job paid well, but I became depressed and developed a cynical view of mankind.

I've had several factory jobs that I soon quit. I rented cars for Budget Rent-a-Car at the airport. This lasted about two months. I was fired because I did nothing while another employee kicked the boss's son a few times, but it wasn't my fight, and I was glad the guy got what was coming.

I remember taking a TV to work to see the moon landing. I don't remember where I worked when that happened. I think this was in Los Angeles. I just could not miss that.

For a while in California I was an orange picker. They would take us to the fields in trucks and we picked oranges all day. They brought us back and gave us $25.

I also worked in a wool factory where I was the only person who spoke English. I didn't like it all that much and I was glad when one day the ax fell. I had enough so I hit the road again.

This story is true; only the names have been changed to protect the guilty.

New York City Phobia

My first trip into New York City with a double trailer rig happened in hunting season. All the northern drivers take off to hunt. I waited for an hour to meet another driver going there. He was a nice guy and agreed to guide me. All the highway signs looked like Greek to me, so I would never have made it without him. The ramps of the big road were built in the 1940s long before double trailer trucks. We had to navigate and keep all the wheels on the road. There were red lights where you could not stop because thugs would rush your truck and probably kill you. It was like one of those zombie movies or like a war zone with burning cars.

Finally, we arrived at the terminal, and it was near the WTC Constantine with a high, metal, electric fence, and security guards in jeeps with automatic weapons. We had to wait for our turn, and the other driver got the first load.

I guess the terror in my eyes begged him to stay, so he did. I followed him back. An hour later we pulled off to a diner. I got his breakfast and then it was a very easy ride back to Carlisle, Pennsylvania. I remember the dispatcher in New York City was dressed like a movie star. He was the "Italian Stallion"; gold rings, tailored suit, and very sharply dressed. He was a very good and helpful person. This experience happened every hunting season, and I soon became used to it - I almost enjoyed it.

My Old Saddle Pal Bob

Some girls from the local college dropped Bob Morris and me off at the interstate ramp near the Tennessee - Virginia border. This was 1969 and Bob had heard I had hitchhiked around the US so he talked me into doing it again. I had a VW Microbus and plenty of money, but when you hitchhike, there is more adventure. I just met old Bob. He was from Washington DC, and he stopped in Bristol to see his cousin who was a friend of mine. Bob was a little skinny hippie that I liked immediately. He woke me in the morning and said, "Let's go to California.". I said, "Let me put my boots on."

Our first ride was from a sailor going back to his base in Memphis. We caught a second ride as soon as we left the sailors car. It was two drug dealers going to Mexico to buy drugs. They had some kind of deal where they would buy an old car, drive it to Mexico, abandoned the car, and fly back to New Jersey with the drugs. I was awakened in the middle of the night to take a pill. I think I was still asleep when I swallowed it. When I awoke, it was daylight, and we were stopped in West Texas. The pill was mescaline, and I was tripping my brains out.

When we reached New Mexico, we stopped at the driver's father's ranch. The two guys stripped naked, and ran with the horses across the desert. Dozens of horses galloping in the sand, and we could see the guy's long hair flowing in the wind as they ran across the desert.

They had left the dashboard glove box open and it was filled with $100 bills. We were sitting there in the car with the motor running to keep it cool with thousands of dollars while our tripping friends ran with the horses. A criminal would see an opportunity here, but we knew they would hunt us down and kill us.

In the middle of the desert, the New Mexico police had established a roadblock. I mean in the middle of nowhere. I was sure they had to be checking for drugs. I knew we were in trouble when the officer said, "Good afternoon, scholars." I was hyperventilating and holding my breath, but after looking at our IDs, they let us go.

When we reached Albuquerque, the guys turned south to go to Tucson, Arizona and then into Mexico. We got out on the west side of town. There were other hitchhikers there. At night in the desert, the air gets cold so we built a fire, and played guitars and shared what food we had in our rucksacks.

Just as a joke, I announced there was going to be some hippies in a bus that were going to pick us up. Dang! Within minutes, an old Greyhound bus pulled over. The bus had been completely rebuilt with living quarters and tapestries on the walls. They even had a new dog on the side. It was two folk singers from Santa Fe who were going to LA to play a gig. Immediately, the girl started cooking. She fed us all. I was telling the driver that I could drive the bus, as I had driven tractor trailers. He was tired, and he watched me drive until he went to bed. I remember holding the throttle to the floor. I was thinking, what an adventure!

The sun was coming up and everyone was asleep so I let that big dog hunt. I was thinking that only a couple of days before I was sitting around watching TV and bored. Now, here I was driving a Greyhound bus across the desert at 80 miles an hour. We stopped the bus in Riverside, California to see a friend of the male folksinger. They were buddies from fighting in Vietnam. We said our goodbyes and headed into LA. That's another story. I remember we had made it to LA from Bristol in 52 hours. A lot of time was saved when we were blasting across the desert, and all of our rides came quickly. This was a road dog's dream if I ever did see one.

Old Love Letters

While unpacking my things at my new apartment last week, I came across a box of my parent's letters to each other during World War II. So much to learn. So much I didn't know about them.

My dad was in the Navy on a merchant ship in 1942. He told me about the ship breaking down in the North Atlantic sea. The German submarines were sinking a lot of supply ships at the time. The fear must've been great for the 19-year-old sailorr sitting there as a duck in the water although he didn't say it. He mostly wrote about when he would get home and see her again. He had two brothers in the Army, but he had no idea where they were. He only knew they were in France

Later in the war he was a gunner on a ship in Russia. He describes how poor the people were, and he had never seen people starve before. He said that if the men at home worked as hard as Russian women, the women back home wouldn't have anything to do.

My mother was working at Oak Ridge for the government helping to build the atom bomb, but of course she didn't know what they were doing. Mostly their letters were about when the war was over and their plans for the future. Love and hope filled the pages. No hopelessness or regret anywhere. Only plans for the time he said he would be freed. These letters showed me the strength of character they had, and how they were able to sustain 62 years of marriage. They built a life to raise my siblings and myself. 75 years later, I am beginning to understand what their love was about. There was no time to look back and feel sorrow or self-pity. There were no regrets the past. It would be over when it was over. I sometimes feel guilty as I think how my life run its course. A lot to learn from old love letters.

Airplanes

All my life I've wanted to fly airplanes. My dad had been in the war, and he loved them. He could look at them, and tell you all about them. I suppose this is where my fascination began.

I was a radio announcer in 1976. A fellow announcer wanted me to work at a station in Bluefield, West Virginia. Part of the deal was he would teach me to fly. He was also a charter pilot for a coal company. This was my introduction to flying. My friend had several airplanes to fly, and he would fly any time to build flight time. We flew two engine planes that were sophisticated and complex. I went to ground school to learn the physics of what was happening and the mechanics of airplanes and avionics.

In 1976, the Mississippi River froze. We flew to Memphis and flew down between the frozen banks only a few feet from the ice. I learned to fly quickly. Nothing is as exhilarating as being up there. One feels sorry for the people making their way at a snail's pace. Eventually, I learned to fly bigger planes, working a few charter flights and giving lessons. The number of flight hours is everything in commercial aviation. By the early 80s, I had a family. At that time, there was a recession, and the only job I could find was in South America. I decided not to go as I had two daughters. So ended my time as a flyer. A wonderful memory.

The sky is such a beautiful environment of color; robin egg blues, angel hair whites, light and dark grades of gray and deep greens. The sky above a bright crystal blue, and land below a green on green checkerboard divided by silver blue ribbons. I remember flying beside clouds that were a mile high. I've seen sights that few others have ever seen. I have flown where the eagles and angels live. Wonder and elation engulf one in this beautiful but hostile world. I was unable to continue my flying, but my fascination with this world will never end.

Live Through This

I will begin this story with the definition of fear. According to the Oxford dictionary, fear is a panic or distress caused by exposure to danger, expectation of pain, dread, terror, horror, fright. Until yesterday this word controlled my life for a few days. Specifically, fear of death to cancer. My history of melanoma is not good; two prior surgeries (the last two years ago). I saw the same look in the doctor's eyes (or so I imagined) that I had seen before the prior surgeries. I knew I was a goner. I looked at the world as how it would go on without me.

I went off the deep end. I had this feeling before the last surgery. Of course, I want to see my grandchildren grow. I want all the pleasures of my life. Things had been going well, and now I had to give it up. It's not fair. I have worked hard all my life and did the right things and played by the rules. I don't deserve to die until I'm very old. Has God forgotten me? Is there a God or is it a tale to use when one is to die? I lost my faith in everything. I thought of all my dead relatives and friends. Two wives and my mother died in my arms. I saw their spirits leave the body. I wasn't thinking of heaven or hell. I was only thinking I was going to die. When I look at my children and grandchildren, I can only wonder how they will do without me. Everything was grim and hopeless.

Yesterday, the test results came back. There was no cancer, only healthy tissue. I am the happiest man that ever lived. Everything is a blessing. Now I am living 110%. To be here and now is everything. No past or future to think about. This morning I hear birds singing and chirping. Yesterday I did not.

The Oxford dictionary defines joy as pleasure, extreme gladness, delight, elation, rapture, ecstasy, happiness. Today, my heart overflows with joy.

Music

When I was very young I sang in church. This is where it started. I would lie in bed at night and listen to WLS, WOWO, WABC, WLAC, and WNOX. I was always singing with the records. When I was nine years old, I began playing the trumpet. I had a great music teacher named Pete Pino. In high school, I played in the marching band and the concert band. Because of Mr. Pino, we had a good sound. I remember how exciting it was to go to the World's Fair in New York and the Winchester, Virginia Apple Blossom Festival, and also the Southeastern Band Festival. We were a small band, working hard on the all the maneuvers for football games. I played all the brass instrument and drums.

In the eighth grade, my buddies, and I put together our own band. My parents bought a set of drum's, and we played Dixieland jazz. After a while, we moved to rock 'n roll. When the Beatles and Rolling Stones hit, that was it. We grew our hair long and played all of their songs. In a few years, we began playing in clubs. We played at the Pink Room in Haysi and all around West Virginia and Virginia. We usually made about $50 each. In the 60s, this was money enough to buy old cars and motorcycles.

The band ended in 1969. I wanted to work in music so I became a radio announcer. I moved around a lot. My radio name was Robert E Lee. When I moved into bigger markets, I made more money, but I had to live in northern cities. Being a country boy, this did not go well with me, and I became tired of the cold weather. Eventually, I gave it up and came home to Bristol, and played in bars again. The name of our band was the Nomads. We replaced some players, played blues, and called it Old South Blues Band.

What a great experience we had with our band, but we did not make much money. The adulation from the audience is why we stayed at it. Maybe this is why all musicians play, but musicians have to eat. Musicians have family responsibilities and so this musician learned to love another sound of music; the rhythms and sounds of the open road. I went from sweet music gigs to those great big rigs with eighteen wheels.

    Once again, many years later, I find myself playing in a five-piece band. We write songs for fun, and as old men we don't care about the money so much - well not as much. It is truly an artistic endeavor. We have good music We are close to having our own sound. If the truth be known, we are only trying to please ourselves.

    I recently moved here from Nashville where I was managing a great singer and band. I learned a lot. It is a ruthless business of rejection. The name of the game is money. That's what it's all about. I got my boys hooked up with Gaylord Entertainment and came home to Bristol.

Reflections of Joe Copley

    I received a phone call a few moments ago that another stroke has taken you from us. In our last conversation, you said you were reflecting. I am reflecting now.

    Remember the 60s when we met. You were too cool in your orange Karma-Ghi, looking like Dustin Hoffman in The Graduate. Everyone loved you. You had it going on. All considered you as more than a friend. They wanted to be your best friend.

    What a wonderful radio personality you were. You influenced all of the wannabe DJs, including me showing us how it is done. Reflecting on our five years working at WFHG when it was a cement building on Valley Drive, you were always giving me concert tickets, dinner tickets and helping me with any project going on. You were a joy to everyone. I was always impressed with your charisma, and with your intelligence and your ease in relationships with others. As a businessman, you were a standup guy, trusted by all and a closer in the art of the deal.

    I have been reflecting on recent times included trips to Morehead city and Cape Hatteras to fish. Images include us riding on the motor scooter in our swimming trunks. After an hour, we had to make it on back home to the blender. The sand and wind had taken off the top layer of our skin. Finally, sweet memories of sitting on a pier as the sun was coming up replenishing the serenity of the moment. Suddenly a school of red snapper comes in and everyone gets busy; Shimbuck, Hopson, Fred Rowland, RJ, the Chamber Brothers, and Glen - old friends making memories bringing in the bounty from the sea.

    I'm missing you already and my world is much smaller without you.

Doc Mongle and Sam

Dr. Mongle had a smile that spoke to you. It was a soulful understanding of the human condition. It was a radiant expression of love of life, including you. People tend to hide negative thoughts about those who have died, but in all the years I knew him, there simply were none. He was truly a wonderful man; a Renaissance man.

Doc and Sam's ancestors settled on the North Fork of the Holston River soon after the Revolutionary war. Two German brothers, veterans of the war, acquired the land and were successful farmers. Doc graduated from Columbia University and started his career at Bristol Medical Hospital in Bristol, Tennessee. In 1952, he built a house in Blountville, Tennessee. I met his son, Sam on the first day of the first grade. We were best friends all of our lives.

Doc was part of the old Virginia bluebloods. I've seen old 8mm films of fox hunting with all the regal splendor one would expect. One film showed Doc at full speed climbing under horse and saddle and coming up the other side. Doc Mongle was a gentry farmer owning over 200 Coon dogs. He had farmhands to look after them. Sometimes, Sam and I would sit in his car with him and listen to the dogs treeing a coon. He didn't smoke, but he loved to dip snuff and chew tobacco. He also liked a little taste of whiskey. He would put in a little water and sugar.

Doc's wife died in 1956 from an aneurism, and he never remarried. Sam had a sister named Bruce. She was 2 years older. She was the prettiest, most popular girl in school. I loved her as did Sam. Many women tried to capture Doc's heart, but he was only interested in his family and his farms. At that time, he owned the old home place in Virginia and a working farm in Blountville, TN. Sam and I did our part at $.50 an hour, and all you could eat at his table anytime.

Larry Carrier and Carl Moore came to him and promised him one third if they could have his farm to build a racetrack.

(Bristol Motor Speedway). He laughed and said that car racing stuff would never last. He might have been thinking he did not want to lose the farm. Doc had a great family, and all the money one could want, but at a great sacrifice for him. I remember his coming home after 12 hours only to return when the phone rang. It seems he had no time to rest.

    Doc played the banjo in an old-time string band. He played the old-time claw hammer style like Ralph Stanley. I remember buses circled near his house when famous bluegrass stars were pickin' banjos in the kitchen. Musicians like Ralph and Carter Stanley, Flatt and Scruggs, Jim and Jesse and many others creating wonderful music at doc's farm.

    Doc was a brilliant surgeon. He served the people of East Tennessee and Southwest Virginia his entire life, saving countless lives. Along with my dad, Doc was my hero. He was a great influence in my life, and I even studied pre-med in college. I learned so much from him and loved him.

    After retiring from the hospital, Doc hung his shingle at his home, and saw dozens of patients daily. He served until his death in the 90s.

    Sam loved my red Corvette and bought one exactly like it. We drove them to the 40th class reunion. When he died, he gave me the car. I still have it.

    Sam's greatest love was his daughter Keri; a gifted child, currently in New York City earning her Doctorate in Anthropology. From the first grade until his death, Sam and I never lost touch. Through hard times, he was my strength. We suffered cancer together. He was alone, and so was I. His diagnosis was before mine so he knew what to say to help me. A very strong man, he lived 11 years with bladder cancer. I was in Big Stone Gap, Virginia when he passed. It would take a huge book to capture our adventures.

    Doc and Sam represented two lives well lived. In the end how could you leave a better legacy?

Cruisin'

I hear some people say how hard they had it while growing up; poverty, hard times, angry parents, drunk daddies, uncaring mothers, repeated hard luck stories from so many people. I almost feel guilty because my life was not like that. My parents worked hard and had plenty of money I was a kid who grew up in the suburbs. Truth is, I had everything. As a teen, I had a job as a porter in the Holiday Inn in Bristol and also played Rock and Roll music. I had my own money and my parents would sign for the bank loans to buy motorbikes and cars. My life in Bristol was like the movie, American Graffiti. Wheels were everything, and we were young and full of energy and passion. Our hearts were alive and free.

Cruising was happening all over the U.S. Man, we loved our cars. Almost all the money went to cars. Every weekend we cruised the local scene. There was Trayers 2, which is now the Hill Billy Market, the Dog and Suds on Euclid Avenue, and of course, State Street from one end to the other. The parking lot at the 'Giant' Food Market' was always a favorite place to go until the cops made you leave. Then there was Carrier's Drive-In and the Blue Circle also.

Sometimes, we would circle cars in the fields where the Wellmont Hospital is now. There was no interstate highway at that time so nobody bothered us. Maybe 20 cars in a circle with all the radios tuned to WFHG AM. We would build a bonfire - maybe 30 or 40 people laughing and having a great time. These were the days before drugs and only a few drank beer. A beautiful star-lit summer night filled with excitement. We had it all. It was a time of innocence before the Vietnam War brought us to reality.

People still cruise in beautiful cars. State Street is more alive than ever. It seems everything has changed, -- and nothing has changed. Bristol will always be my home, and if I die in Bristol, at least, I will die free.

Sergeant Pepper's Merry Pranksters

The winter of 1963 was unusually tough; long days in school and little else to do. A plan for Spring Break was hatched between 6 of us, all 15 years old. It had to involve a Myrtle Beach road trip. Our friend John's dad had an old laundry truck that he had made into a camper. Her said we could borrow it if I did the driving. My birthday was June 7th and by noon I had my driver's license.

We left Mayberry (Blountville, TN) at 2:00 PM. Most of us had robbed our Mom's cupboards of canned goods and plenty of green beans. Dave had gotten a case of Diet-Rite Orange soda. It tastes like fly spray. We drove to Nag's Head, N.C. and followed the Grand Banks. This is something everyone should do. At Nag's Head, you can walk up the hill where the Wright Brothers flew their first airplane. Incredible beauty; the lighthouse at Cape Hatteras, wild horses on Ocracoke Island, that no one has any idea of how they got there, and Bird Island where there is nothing but birds of every description. You see it from a free ferry ride from the state of North Carolina. We all slept in the camper on the beach. Our next destination was Myrtle Beach.

We found a place in the campground and headed for the strip. We went to the boardwalk and drank beer even though no one had an ID. Of course, we were all on the make. Girls were everything at that age, and we all thought we were ladies' men. We stayed out late and when we returned to the van, we found we had forgotten to lock the door because we were dumb kids from Mayberry (Blountville) and everything of value had been stolen. We learned the hard way that the world away from home was a very different place indeed.

Some of our buddies had a hotel room on the beach. They were a group that I played music with. The most important things in our lives were girls, music, motorcycles and fast cars. The guitar player in the band had brought his record player and a copy of the new Beatles album, 'Sergeant Pepper's Lonely Hearts Band' and it changed my life. This was serious music with a serious message. I had never heard anything like it. It was transcendental; beyond my musical experience and knowledge. It changed the way I listened to music and played.

One morning at our campsite, a Dunkin' Doughnut truck came through. Everyone was asleep until we heard the doughnut truck music. We were getting out boots on to eat doughnuts when we heard Wayne tell the driver that we didn't want any doughnuts, Man, everyone jumped on him for that, and we still kid him to this day about 'Dunkin' Doughnuts".

We finally made our way home. We arrived at sunrise, and I remember Wayne looked around and said, "It's changed." Of course, we all got a laugh, but the truth was that we had changed. That was the first time we had seen the world that wasn't from our Dad's back seat. We had taken a huge step in growing up. We had learned a lot about the world away from our little town. As Earnest T. Bass would say, "We had seen what was beyond Kelsey's Woods".

Flying to Knoxville 1988

I had rented airplanes at Tri-City Aviation before. This meant a check ride with an instructor as you circle the airport and land. The plane was mine for three hours' flight time at $25.00 an hour. Pre-flight involves 21 steps such as draining a small amount of fuel from the bottom of the tank to look for water plus twenty other steps. When you are certain of the airplane's safety, you call taxi control on one of your 2 radios. You tell them where you are at and they give you the number of the taxi-way to the run up area. This is a big circle where all of the airplanes do the final check list. They tell you the wind direction and you turn your airplane into the wind and set your altitude meter and your DG which is your guidance system. When you are ready to go, you call the tower. If there is no traffic, they give you the active and that means go. Don't fool around. It is time to fly the plane. Push the throttle against the firewall and steer with your feet. When the airplane reaches 55 miles per hour, pull back on the yoke and you fly. Your best rate of climb speed is 70 mph and so you maintain 70 mph by raising or lowering the nose. Raise the nose and it slows down and lower it to go faster. While all of this is going on, you must dial the radio to departure control. You tell them where you want to go. They tell you the heading and assign you an altitude to fly. After this, you are off radio and maintain altitude and heading. At 25 miles away from Knoxville Airport, you call them on approach control and they might already have you on radar. If they don't, they will ask you to push a button on the transponder which makes your spot on the Knoxville radar glow brighter and bigger. They assign you a new heading and tell you which runway number you are to land on. Then you call tower and they give you a wind read on runway. When you touch down, you call taxi control and tell them where you want to go. They assign you a taxi-way and you have arrived. To go home, reverse everything I have told you.

Duct Tape and Onions

I had been in Potacala, Idaho for two days. It is an incredibly beautiful place where the air is so clean and clear you can almost touch the jagged mountains. It's a great place. I was leaving, having found a load in Oregon. I drove to the fields and the migrants loaded my trailer with 48,000 pounds of onions. It was October and the air was crisp; almost stinging my lungs as I left for Atlanta.

Way up north on I - 90, making my way to Cheyenne, Wyoming, the temperature was dropping quickly. I stopped at a Wal-Mart and bought insulated coveralls and duct tape. I got some food then used the duct tape to seal every door, vent and window in the truck.

Out of nowhere, a blizzard hit; 50 mph winds full of snow on the north slope of the Rockies. The interstate shut down. On the map, I found a little road going up the mountain. If I could reach the top, it was all downhill to Denver. I could not see the road, but I rolled down the right window and I could see the guardrail. I stayed away about ten feet and was making my way to the top. On a particularly steep grade, the truck came to a stop. That was it. I was sitting on the north slope of the Rockies and had no heaters to save the freezing onions. There would be a lawsuit and I would end up paying for the onions.

I was sitting there pondering bankruptcy when I saw lights coming from behind. It was the Wyoming State Highway Department. They had a huge machine and they pushed me to the top of the mountain. Rolling down the southern slope, I was in a clear, star-lit night. I was amazed to think what was going on a few miles behind me.

I made it to Atlanta ok and had to pay $50.00 to enter the produce area to deliver their produce to them. Almost always, the urgency concerned me as it was produce. They were overcrowded, and they decided to sell the onions off my truck. It took three days.

I called the ICC, and they told me the produce guy had three days to sell it. The laws are totally against the independent trucker. I was so disgusted that I brought the rig back to Kingsport. The only consolation was that the loaders in Oregon put fifty bags of onions too many in my truck. Everyone that I knew got plenty free onions. Worked like a mule, worried like death row and lost money. Independent trucking is feathers today and feathers tomorrow.

Truckin' with Earl

    I was fourteen years old when my best friend Dave's older brother drove a truck. It was in the summer and, I was waiting near the highway to leave on my first big truck ride. It was a very old rig; 1953 Mac H Model. When we reached the interstate, Earl had taught me how to shift gears, and he climbed into the sleeper and went to sleep. I managed to get the rig to Rockwood, Tennessee where the interstate ended. You had to get back on old highway 11 and it was a dangerous two lanes through the mountains with steep drop offs and steep hills. The truck also had two gear shift levers and was complicated. This was my first real test. This was an old world rig with old world brakes and power. I drove the truck across the Mississippi River at Memphis. Earl was well rested, and I had made the grade.
    Yesterday, I could not even spell truck driver and now I are one. It doesn't take long to get to climb the ladder in this professional career. We delivered the load in McAllister, Texas.
    The truck was so old, there was no air conditioning. The temperature was about 110 degrees in the truck. There was so much smoke that we had raccoon eyes when we got out of the truck. We headed back to Arkansas to load bags of rice headed for New York City. I remember driving through the Holland Tunnel and the. exhaust stacks almost touched the tunnel. We delivered the rice to merchants door to door. The people who were unloading the truck were speaking Spanish, and we couldn't understand a word. We knew they were laughing at us and cursing us because we were hillbillies. As far as I know, there is only one truck stop in New York City, and it is called Toolies. After a couple days, we had unloaded and loaded munitions housings bound for Texas.
    I remember going into a bar in New York City with Earl. The bartender ask me what I wanted and I said, "Vodka martini - shaken not stirred." because that is the name of the only drink I knew of. The bartender did a little extra shaking as he handed me

my drink with an unpleasant smile. It was my first drink ever and it was the worst tasting drink ever.

There were prostitutes everywhere, but they didn't talk to me because I was a kid. We made our way to Texas and unloaded. I remember one of the men unloading the truck was an albino. Having never seen an albino, I was amazed at seeing his pink eyes.

We went back to Louisiana to pick up more rice when the truck broke down. When they jacked the cab up to fix the motor, the cab which was located over the motor fell off into the street. A wrecker was called and the cab was placed back on the truck. We loaded rice for New York again. When we got home, I had seen the USA and what was out there. To a Blountville boy, Texas was like being on the moon.

I had an old car, and I drove it to see my buddies. I had on a cowboy hat and was smoking a big cigar. My buddies all called me Fidel. My first truck ride; a ride that I will never forget. I will never forget, "Truckin' with Earl".

Forty Below

As I arrived at the terminal in Rockford, Illinois, I entered a surreal world of frozen snow so hard that the wheels of my forty-ton rig would not break through. Everything was so quiet, with all sound muffled by snow. It was like a scene from Dr. Zhivago. At this temperature, you have to put additives in your fuel to keep it from freezing. If your motor quits, you won't last long; you may be a dead man. The air was so cold it hurt your uncovered face, and even breathing hurt.

I reached the deserted truck terminal and called the hotel. They told me there was a curfew, and it was illegal to be outside. They refused to come and get me, and I was to sleep on the floor. I called my company home office to report my situation. In twenty minutes the hotel van showed up. Thank you Jimmy Hoffa.

When I reached my room, I turned the heater wide open and turned on the TV. Dan Rather said Rockford, Illinois was the coldest place in the country; exactly where I was. I removed the curtains from the windows to cover me in bed. This room was not built for forty below. My delivery was to the Chrysler Plant in Rockford.

I looked over the factory while I was there and was amazed at the technology at that time. Just imagine what it is now. This is where Chrysler made the Dodge Omni, one of the worst cars ever. People began driving Japanese cars and one of the reasons was the Dodge Omni.

When it was time to leave, I was glad to have a warm heater in the truck. I have never been that cold for so long. This was the big winter 1978 and I never saw pavement for months; only white roads. At times, the snow was piled so high beside the road, it was like driving through a tunnel. The snow was piled 25 - 30 feet high on both sides of the road. I drove from Detroit to Chicago and never did see the sun.

Big rigs are heavy and go well on the snow, but they don't

stop very well. It is hard to get 40 tons to stop quickly. In fact, it hard to do anything at all. It's like you are a passenger with occasional voting rights. During November of 1978, I left Charleston, W. VA at midnight with no snow. About 2:00 AM, I was creeping along on a two lane road with other big trucks about 15 miles per hour trying to see the taillights in front of me. The rig in front of me was blown across the road and crashed. The trailer covered the road completely, and so there we were. The winds were 50 to 70 miles an hour loaded with snow.

There were no cell phones in those days and so all I could do was sit in the truck. In a few hours, a guy in a wrecker shows up and says the entire state was closed down and I was to go to his house. I climbed into the wrecker and when we had arrived at his house, he had already brought five other truckers.

For three days, we played cards and watched the weather on TV. The hostess was a great cook. At the end of three days, the governor of Ohio announced that the National Guard was going to open the roads. Everyone had 24 hours to move their vehicles or the state would move them with a fork lift. It worked and everything went back to normal. With overtime, I made over $1,000.00 for the week: the very first time ever. I have never known forty below again.

"Race" Weekend

Our camping site was beside a beautiful $350,000 motor home. My family made friends with the sole occupant. He was my age and had quarreled with his wife. He had recently sold his business and retired. He was supposed to be having the good time that wealth and no responsibilities bring you, but it was not. He was unhappy. He was an old graybeard like me, and we became friends. His wife had rented a car and was gone so he needed a friend to hang out with. We drank too much, and he fell down on the concrete. The others around us were 99% college kids, and they picked him up and gave me the things he had dropped. These young kids were treating him with respect and kindness. They were hugging him and making sure he was taking care of. I was impressed by the respect shown to us old graybeards.

We were having a good time, but my friend had to go lay down. After the concert, my family had fixed some great food. The grandchildren were having a great time. My new friend was the only person I saw that was visibly drunk. It's more of a family thing now. Everyone smiling and having a good time. My new friend told me his life story over the weekend as people tend to do He came from blue-collar and started a construction company. He was a multi-millionaire who had lost two children, and was about to lose his last son to drugs. His wife was unhappy even though she had four houses to live in when she wanted.

He was amazed that I was happy, and I ain't got doodley squat. My wealth is in my family and friends. I have the arts to keep me entertained. I would never trade places with my new friend. He feels like his life was wasted making money as our consumer culture dictates and he was wrong. During my life I have known people of all economic groups and most times the wealthy are the unhappiest. All too often they didn't own the money. The money owns them. Money can't bring you happiness, but it can buy a big boat and float up close to it. Once again the facts may have been changed to protect the guilty.

Troubles

I heard an old-time preacher say, "If you don't have any troubles at your house just wait, it's coming". All my troubles have overwhelmed me. I have a child on drugs. All she wants is to drop out and not participate in life. This has to be a living hell.
I have an enlarged aorta that could rupture and burst anytime. My brother-in-law and dear friend for over 40 years died two days ago. Two days ago, I crushed the fender on my car which I just bought in the past month. Except for the crushed fender, I turn everything over to God.
Our time here is short and full of trouble. I've tried to live without God. It didn't work. Only God can help me find a life that is wonderful. Every day is full of adventure. Strengthen the ties that remain. Tell those you love that you love them. Go for a walk in the park or anywhere. See the beauty and life around you.
I have friends to help me. Dear friends that I can't do without. Fences define neighbors, but no distances can keep friends apart. I have heard that friends are God's way of apologizing for families.
Enjoy this day and participate. Pray without ceasing. There are more pleasures in life than troubles. Whatever it is that you are doing that you enjoy, do it and do it often.

Fast Ride

      Blast out of the turn with your body hanging off the bike in the direction of the turn, and your knee is almost on the ground. When the bike gets full up, slam the throttle wide open until the next turn in only a few seconds. You know the fast way through the curves and you lay out the line you will follow. You hang your body off the bike using only your front brake to slow the bike. Keep your foot on the back brake lightly. This helps to keep the bike steady. It is called trail braking. Hang off the bike into the turn pulling the handle bars in the opposite direction of the turn. This bends the bike down into the turn. This is not so hard unless you are racing other bikes. The adrenaline and the excitement is so intense you sometimes forget to breathe and you have to remind yourself. I am talking about 100 miles an hour and more. It takes nerve plus a lot of skill.
      When you are drag racing, you do what is called a hop. You must have the bike completely straight up. Rev the motor to 5000 RPM and dump the clutch. If the bike is not straight up, it will not go straight. After a couple seconds, you change gears. Once again, you must tell yourself to breath because of the intensity. Speed shift without letting up on the throttle. If everything goes right, it doesn't scare you, but your heart is pumping hard.
      At extreme speeds the entire world becomes a blur of color. There is only one spot of clarity that seems like the size of a quarter. You dare not move your eyes off that spot. I can't say fast riding is safe, but there are ways to reduce the risk. The bikes are built for this and you ride head first with your rear in the air. This puts more weight on the front wheel and creates more control. Buy the best tires you can buy. Only a couple of small spots arte touching the road, and you have to trust your tires and know the limits of their grip on the road. Racing tires are very soft to grip the road and wear out about every 2500 to 3000 miles. These bikes

require maintenance. One must tighten the chain and have the right air pressure in the tires. Always wear leather clothes. If you wear street clothes, you will tumble if you fall. Leather cloths help you slide. Of course, leather won't help if you slide into a guard rail or over a cliff. Body guard equipment such as fiberglass strips to protect the spine also helps, and wear elbow and knee protectors.

I cannot explain my burning need for speed, and I cannot tell you why I would do it again this morning.

Time

My friend Joe told me in our last conversation ever that he was reflecting. "How did all the time go by so quickly?" I remember all of it well. I heard a very old man say, "You try something you like until you're bored of it, and you go on to something else." Literally, you're bored of everything so you just sit around. Sometimes, it is hard to have the energy and desire to do anything. How many people are hiding behind locked doors doing nothing? Sitting in easy chairs will kill you sure as cancer. I know people in their 50s whose entire day is about what is on TV. On a beautiful day they're watching TV.

In life everything is new. It took the long and winding road to get here. Here and now is all we've got so let's celebrate. May 14, 2015, the first day of the rest of our lives. I have a bucket list. It involves seeing a couple of NASCAR races, racing through the Outer Banks again and perhaps a new motorcycle. If you have a passion, that in itself is key; then pursue it. Time is short. I'm in a hurry. Every day, I must first enjoy the day and also get something done. Getting older isn't for sissies, but I recognize it and accept it. My heart is young and free. Me and time, that's where we are today.

Coconuts and Mustard

This adventure began when some of my buddies moved to Florida. This was in 1968. I received a call to come on down. I told my friends and a road trip was born. I had a VW Bus that the girls at VI College named 'Biesel'. Four buddies and myself went to Jacksonville. When we found our friends, there were already four guests in his apartment and so there was hardly room to stand. Also, the 'vibes' weren't good because of some drama between my friend and his girlfriend so this was not the place to stay. He told us about a park named Coconut Grove where we could park 'Biesel', hang out at the beach, and enjoy Florida.

There was a restaurant across the street, and they had live bands. This went well for a few days until we ran out of money. All we had to eat was coconuts. Someone found a jar of mustard in my van so we ate coconuts dipped in mustard. Every one of us had diarrhea. The police came and told us to go away. We had made friends with some of the locals. There was this kid we called Sal Mineo because he looked like the movie star. He had a small dog. He became one of the guys and he told us about a place where we could stay. We drove to a house in Jacksonville. I had called a friend in Bristol to send money, but he had to wait until payday. Which was a few days away.

Our new place had a major problem. There were five 'Hells Angels' staying there. These were some bad guys who had everyone terrorized. These guys would fight with each other and one night one was thrown out of the window on the second floor. These guys had been arrested when they came from California and the police had taken their bikes, I wasn't afraid and they treated me with respect. I had a guitar and I played music for them. They could tell who was afraid and picked on them. Our new friend Sal Mineo was stealing food from grocery stores to feed us. He would show up with luncheon meat and bread.

After about four days, he found us a new place to stay; the

city dump. I imagined a stinking mess, but it was only trees. People had dragged couches and chairs, and even had a TV in a tree even though there was no electricity. It was like sitting in your living room – in the woods. Every night, people would bring instruments and we would play by a fire.

On New Year's day, January first, we were busted by the cops. The harmonica player had pot so he was arrested. I remember the cops asking if anyone knew how to spell marijuana. They had us follow the cruiser to the jail and even told us to stay in the parking lot. The police told us the jail was full and we would have to leave the county. They followed us to the county line and turned around. My friend had sent us some money and we picked it up at the Western Union and came on home to Bristol. As always, the facts of this story have been changed to protect the guilty.

High Water

I remember driving through Times Square in 1971. I was driving a 1956 Chrysler with a 2 axle and U-Haul behind. In the car was my new wife Judy and her two daughters who were two and four years old plus her mother and two small dogs. In the U-Haul was everything we owned. We had decided to move to New York City. We had been living in Frederick, Maryland. Some of my friends and I had leased a hotel and restaurant. We had previously traveled to Washington, D.C. and put an ad in the newspaper looking for a place to lease.

We got a call from Nicolas Basilico who owned a hotel. The facility was called the "Princess Inn" and we all moved there. There were about ten of us. We had gotten together about $5,000.00, and we all had jobs in the hotel. Our plans involved leasing buses to bring people to the hotel. Area sports were blacked out on TV and we thought we would bring people to the games and weekend for a cheap price. We had plans to hire down and out people; a very good idea on paper, but it didn't work. After about six months, I gave up. We loaded up and left for New York City. We had enough of Washington, D.C.

I saw NYC as the center of culture and opportunity, but circumstances ($1200 a month rent and no job) killed this dream so we agreed to return to Tennessee. We moved to Knoxville. I bought a 70-foot trailer. On the 2nd day, I went to the biggest radio station in town and asked for as job. They gave me an audition as a radio announcer. They had just lost the DJ on the all-night show. They asked me to start at midnight. On the way home, I pulled the car over and danced around because I was stunned at my luck.

In a few weeks, I bought a new Ford station wagon. We stayed in Knoxville three years and finally moved back to Bristol. We moved into a house on South Holston Lake. I worked for WFHG which was the most popular at the time. From 10:00 AM until 3:00 PM, I was the DJ; working five hours a day and making good money and living in the wilderness at the lake was wonderful.

    We lived there four years and decided to try our luck in northern cities where there was more money. We went to Charleston, West Virginia for a couple of years and on to Columbus, Ohio, and Chicago, Illinoi. After this, we were sick of the northern cities and came home again. I worked at radio and played music at night. Everything the north gives you, there is a high price in return: unfriendly people, crime, hatred of the blood, and hatred of 'hillbillies', as we were, and we hated it. That is how we felt then. There is no place like home and this story is about learning this. This is part of my story and has led me to the here and now Hard times make you tough and make you appreciate the many good things in your life. Things are tough out there---- High Water everywhere.

Jess and Alma

Jess and Alma Estes were my grandparents. They had nine children during the depression. One child died at birth and one lived two weeks. All were born at home. They had 50 acres of land, and he was part-time farmer, but he had a job at Oak Ridge. During the depression, he walked 8 miles to catch a truck that would haul him to the Smokey Mountains National Park. He worked for the W PA, and they built the highway across the mountains. When he came home, he had to walk the 8 miles home and then milk and feed. He was paid one dollar per day.

There were some very hard times and my grandfather used to stop at a huge tree near the farmhouse where he prayed every day. Many years later, they realized she was going every day to pray at the same place. Neither knew that the other was doing this. They lived in the country near Sweetwater Tennessee in a community named Sunnyside. All dirt roads and everybody knew everybody. In a way, they policed themselves. If a man was too lazy to work and take care of his family, the other men in the neighborhood would visit him at once. They told him if he didn't get a job they were coming back to beat him until he found one.

I remember the peddler was an old truck filled with everything you could imagine from candy to farm tools. My grandmother would give the me eggs to get candy as eggs were like cash to the peddler. My grandmother would buy bolt cloth. Every Saturday we would go to town to shop and have a great time. Sweetwater was filled with farmers sitting around swapping pocket knives, chewing tobacco and telling lies.

My great-grandfather was a veteran of the Civil War. He and eight of his buddies walked to Kentucky to join the union forces. During the night, they walked into the Confederate Army. The rebels wanted to know where they were going, and they told them they had come to join the Confederate Army. They were told to find a place to sleep and they would be sworn in in in the morning. In the night, they slipped off and continued to Kentucky to join the union. They were poor country boys, strong Baptists, and had no slaves. He fought at the battle of Franklin and at Shiloh. These were killing fields and thousands died.

After the war, he came home to Sweetwater and opened a blacksmith shop. He had seven sons and they prospered. He opened the first Mason Lodge around. My grandfather was the seventh son of the seventh son. My grandfather was born in 1896. He was not in the first world war, but some of his brothers were. To supplement their income my grandmother worked in a weaving mill. My great uncle Harris was sort of an eccentric old guy. He had a beautiful house in town where he lived with two unmarried old maids and was the talk of the town. He looked exactly like my grandfather, and we visited him every Saturday. These were strong religious folks respected by all. All a man has is his reputation and these are the finest of men.

Every Sunday we would put on our best to go to church with dinner on the ground. Wonderful memories of the people I love. At nine years old, they let me drive the car to church. I have their picture on my wall and always smile as I look at it. Now I'm a grandfather, and he taught me how to have unconditional love, like my grandparents. When they retired, they bought a motel. It was a large old time cabin on the main highway. He wanted to watch the time go by and make many friends. When they got too old, they bought a great house and remained there until their deaths. He died at 94 years old and had never smoked a cigarette or tasted alcohol. I remember on his 66th birthday he had read the Bible 66 times. Thank you God for these wonderful people who loved me as I love my children and grandchildren. I understand their pride and their happiness.

A Story About Life at Another Time

(as told to the author by John Ed Robbins who was like a second father)

  In November 1945, John Ed Robbins been floating in a landing craft off the shores of Guam Island in the South Pacific for three hours. The gathering of the invading force had taken three weeks, and he was eager to get off the water. The floor of the vessel was covered in vomit as some of the soldiers had eaten a hearty meal thinking it would be a long time before they ate again. Three marine divisions of the Army 77th Division were invading, and the gathering was slow because of the human factor. Unforeseen problems were slowing the process. It was late in the afternoon and the 305th brigade were waiting their turn. Many were worried about landing in the dark.
  For many of the soldiers, it was a time of reflection. John Ed was thinking about Panama where he had left to invade Guam. They were on high alert as they did not know if there were Japanese submarines off the coast to invade. He was standing guard duty at 2 AM. Something startled the howler monkeys and John Ed opened up with his machine gun. At dawn, there were dead monkeys everywhere. Nothing was put on his record, that his Sgt. had a field day with him.
  He chuckled out loud as he thought of Needmore, Virginia. He remembered driving to Kingsport to pick up his sister. He had no license, but had been driving since he was 10 years old. He drove his sister to their uncle Sid's house in Norton, Virginia. A street light was shining a half block away as they got out of the car to greet uncle Sid. Suddenly, a ghost like object came flying at them. Uncle Sid screamed, ran into the house, and locked the door. John Ed and his sister quickly dove into the Model-A Ford and left wheels spinning, but the ghost followed them. John Ed didn't believe in ghosts, but there it was in the rearview mirror. Suddenly

it was gone and the drive to Gobblers Knob was uneventful.

In the morning, on his drive back to Norton, John Ed saw a white kite hanging from an electrical wire. This was the ghost which had been tangled on his rear bumper -mystery solved. John Ed was a tough kid. When he was 12 years old, he and a friend walked for two days from St. Charles to Needmore; a distance of 62 miles. John Ed was lost with sweet memories of growing up in Southwest Virginia.

The Sergeant was yelling to get ready bringing John Ed back to reality. He was worried about the delay and his hand was hurting. When departing the troop ship, climbing down the rope ladder, a soldier above had stepped on it.

The invasion was behind schedule because of slow going in the invasion of Saipan three days earlier. The purpose was to weaken Japanese defenses for the invasion of Guam. Guam was 34 miles long and would provide a bomber base for the invasion of Japan. Tensions were high in the 305th as they departed the landing craft. The men had to cross a 500-yard coral reef under fire. Burdened with hundreds of pounds of equipment, steel helmet, life belt, rifle and bayonet, grenade launcher, ammunition, rations, a pouch of hand grenades, two canteens of water, and John Ed also carried heavy machine gun parts. He struggled for footing on the coral sometimes neck deep in water. Some men were attacked by sharks and some simply drowned. When John Ed reached the beach, he was a fighting S-O-B, he was there to stay. He was not going back into that water.

The marines had established a beachhead. By midnight, an all-around defense was established, and the 305th dug in. Two battalions were on shore and another on the way.

During the night, several attacks were made on the GIs, but all were repelled. Seven soldiers died, some possibly by friendly fire. The enemy had surrounded them on
three sides and controlled the high ground. The objective was to circle the high ground and push the enemy south to the waiting

305th.
    John Ed had set his machine gun to meet the enemy fleeing south. Hundreds of Japanese were killed. He fired his gun until the barrel turned red hot. With asbestos gloves, he replaced it and continued killing the enemy. Slowly, the high ground was captured. The southern half of the island was secured, but the northern half was the prize with its airfield and installations. The terrain was impossible and the fighting was intense.
    Soon, John Ed found himself scaling up steep slopes and sliding down narrow ravines. The equipment on his back was hanging up on the jungle foliage. Heavy rains created a quagmire. During the night when the soldiers needed warmth, they sat together in flooded foxholes with their teeth chattering. During the day, their mud encrusted uniforms were wet and sweaty. Mosquitoes tormented them at night, and the flies took over at dawn. There was always the mud. Helmets, uniforms, equipment, and even their skin took on the color of Guam soil - a dirty red. The vegetation was so thick advancing columns became lost. Beaches were covered with dead bodies from both sides mixed up together in a cherry red meat pie black with flies.
    By August 6th, the Infantry was in position to continue the attack. At least, 2000 enemy were still effective and well dug in. Enemy forces believed they, and their families, would be forever disgraced by surrender or capture, and could never return to their homeland. In this hopeless situation, many killed themselves by hand grenades or simply waited for death which was not long in coming.
    Enemy tanks were destroyed by bazooka rockets. Japanese had to be forced from caves with flamethrowers. Many of the enemy had taken shelter in huts along the trail just waiting to be killed.
    By August 8th, the 77th had pushed the enemy to within 1000 yards of the sea. The northern island had to be mopped up. The 77th made attempts to induce the enemy to surrender by

dropping leaflets with no results.

John Ed was ordered to the cave area. The patrol encountered ambushes at every turn. There were many caves and well-worn footpaths in that volcanic landscape. The 305th closed many caves, but could not reach the tunnels. The fighting was close and intense. The soldiers withdrew at dusk having lost 8 who were killed and 17 were wounded.

At dawn, the 305th blasted away the tunnel entrance. Some of the enemy ran out and were killed. Before the Americans blew up the tunnel, the Japanese inside were heard singing some sort of chant. The blast sealed the cave. When it was opened four days later, the GIs had to wear gasmasks because of the odor. Sixty dead Japanese were found inside and 300 were killed in this operation.

On August 10th, all organized resistance on Guam ceased. Garrison forces and Seabees began building runways for the new B-29s. Natives were returning to their villages and farms. They returned to rebuild their devastated homes.

On November 3rd 1944, 18,000 men of the 77th Division loaded aboard a squadron of transports in Guam headed for New Caledonia for a much needed rest. The officers and men were to go to a peaceful island free of Japanese. They were catching up on their sleep, relaxing from the tensions, the horrors of combat and enjoying fresh food that never tasted so great. Eight days into the voyage and four days from New Caledonia, radio orders arrived sending the convoy to Leyte, Philippine Islands. The war needed the 77th as things were going poorly.

The convoy arrived November 23, 1944 to a welcome rain. The enemy airplanes could not fly in the rain. The east coast of the island had been taken by the 6th Army. The 77th was not expected to enter combat immediately, but that did not happen. Heavy rains stuck the artillery in the mud, but they would provide fire for the 305th. Natives were rounded up to carry supplies as there were no docks. The island consisted of finger ridges and deep ravines with nearly perpendicular walls. The wounded had to be carried two miles. No jeep could reach the front.

John Ed's company was ordered to protect a Navy airfield constructed on the island. On the night of November 24th, the Japanese sent four plane loads of specialty trained paratrooper raiders to fight the 305th. The fighting was fierce, but the airfield was protected.

The 77th Division was ordered around the island to establish a beachhead. Every landing ship was beyond capacity. All the soldiers on both sides were scared, but as seasoned veterans, they knew it did no good to think of what was coming.

The Allied Forces encountered only light resistance which allowed the 77th to advance rapidly. As they moved inland, the fighting increased. The plan of waiting and meeting them inland had worked well for the Japanese in the invasion of Tarawa where they had killed 5000 Marines.

The Japanese had machine gun emplacements under the native huts and had to be routed out hut to hut. By 1600 hrs. the 305th and captured the village. Much war material was captured and destroyed, but trucks were pressed into service immediately. Orders were to consolidate for the night and prepare for the enemy machine gun, mortars, and rifle fire. The approaches were open and the devoid of cover. The advance halted. Men of the 77th cradled the machine guns in their arms and the advance continued. The men call this the, Battle for Bloody Hill.

John Ed came upon a Japanese corpse and stripped it of a samurai sword. Some were valuable. He carried it for a few days, then threw it to the ground. He was tired of carrying the extra weight. Another soldier had seen him throw it down and rushed to pick it up. He only carried it a few days and threw it down for another soldier to find this war 'booty'.

After the capture of Ormac, the job was to clean every inch of the island of Japanese. John Ed's platoon came upon an island atoll. He and another soldier were ordered to climb the atoll and look around. Walking out of a grove of palm trees on to a white beach, they surprised 100 enemy soldiers on the beach. They were looking at each other eye to eye. They had been waiting to see an American soldier to blow themselves up with dynamite. John Ed was dumbfounded, but he had no time to understand why. He was glad because now he did not have to kill them. The fighting moved on.

Two sides were looking at each other with different world views: The Americans were fighting to get this over with and go home to loved ones, cornbread, buttermilk, and tomato sandwiches. The Japanese were trapped into a belief system that was thousands of years old. This forced them to blow themselves to kingdom come for the Emperor. Eye to eye contact; there had been no compassion on either side. They may have just as well been looking into the eyes of a goat. Neither side saw humans.

The Japanese had to be driven from house to house, fighting with bayonet, mortars, by hip shooting riflemen. The 305th was met with rifle and machine gun fire. Mortars and self-propelled weapons kept a constant barrage in front of the 305th. Cooks, drivers, mechanics, and clerks were deployed and advanced under heavy fire. Japanese soldiers were hiding in 'spider holes' they had dug and covered with coconut logs and brush. John Ed had to be within ten feet to find one. Hand to hand and point blank firing, hip shooting and hand grenades - as tough as it gets. The heat and humidity of the day combined with uphill fighting caused scores of soldiers to fall out. They were evacuated to aid stations, splashed with cold water, and laid in the shade. The fighting men doggedly returned to combat. Only steadfast determination saved the day.

Between December 7th and February 5th, the 77th Division captured Ormac Airfield. This marked the end of the struggle for Leyte. 19,456 Japanese had been killed or taken their own lives. 543 Americans had been killed. For every American death, 36 Japanese died.

John Ed was in eleven island invasions before his time was

up, and he came home. As he was leaving, he noticed a soldier staring at him. He recalled the man had given him a hard time when told John Ed was short time. Because he was so young, the soldier thought he was lying. John Ed remembered waving goodbye to this astonished soldier.

A story of another time about which very few will ever hear. John Ed Robbins died at ninety-one, a few weeks after sharing this story with me.

Made in the USA
Middletown, DE
06 June 2023

32154162R00051